D0140067

Chemistry
and
Artists' Colors

Mary Virginia Orna
and
Madeline P. Goodstein

Publisher:
ChemSource, Inc.
www.thenewchemsource.org

Copyright © 2017 by Mary Virginia Orna

All rights reserved. This book may not be reproduced in whole or in part, stored in a retrieval system, or transmitted in any form or by any means – electronic, mechanical, or other – without written permission of the publisher, except by a reviewer, who may quote brief passages in a review.

The material in this book is intended for educational purposes only. No impressed or implied guarantee as to the effects of the use of the miniexperiments can be given nor any liability taken. The publisher advises that the activities contained herein not be attempted or practiced without appropriate guidance or support.

ISBN 978-0-9637747-6-7
Cover: Olivia Thomson (oliviathomsondesign@gmail.com)
        Indiana Wesleyan University

Table of Contents

Part II. Matter and Color

Part III. Chemistry and Color

Part I

Introduction
to
Light and Color

Science is nothing else than the
search to discover unity in the
wild variety of nature – or more
exactly, in the variety of our own
experience. Poetry, paintings, the arts
are the same search, in Coleridge's
phrase, for unity and variety.

Jacob Bronowski
"Science and Human Values"
New York: Harper, p. 16 (1965)

CHAPTER 1
INTRODUCTION

While art has always been with us, from the earliest cave paintings right down to the most modern free-form sculptures, the science that allows the artist to control the use of materials, chemistry, is still in its infancy. The artist has always had to deal with materials and objects in order to explore expression. Chemistry is the science that deals with materials and the changes that take place in them. It is the science that is concerned with discovering something about reality by an examination of matter itself. Art uses matter, in the form of materials and objects, in order to interpret the artist's view of the universe. Both disciplines are very closely tied to matter, but in very different ways.

One of the properties of matter that intensely interests both artists and chemists is color. Color is an invaluable vehicle for the artist; it is often an analytical tool for the chemist and has opened the door to an immense realm of chemical theory.

Until about a century ago, it was necessary for artists, or at least painters, to have the skills of manufacturing chemists in order to provide and to control their own colors. The commercially available artists' paints, media and solvents that we take for granted today were unknown until well into the nineteenth century. It was sometimes necessary to synthesize the pigments, or grind raw minerals to the proper particle size, and intimately mix them with vehicles in order for an artist to control the grade of the final product. This fact becomes obvious when one examines the painters' manuals of ancient and medieval times. The bulk of the instruction in these works is concerned with how to manufacture the pigments for painting. The ancients were skilled at marshaling almost every colored substance in the plant, animal and mineral worlds for this purpose.

Although today's artists have the finest grades of both synthetic and natural materials at their fingertips (for a price!), such availability can sometimes serve to remove the artist from the level of concern about the nature and properties of art materials. The Old Masters had an intimate knowledge of their materials because they often manufactured their own from very crude starting materials. Today's artists need only squeeze a tube in order to have a high-grade, homogeneous oil paint on the palette. The steps from starting materials to palette are a mystery to many an artist, and some think that it is better that the mystery remain untouched. Other artists feel the need to know more about the materials they use, their properties, their modes of interaction with one another, their stability to light, heat and air pollutants. Still others would like to know more about how their materials interact with light, why colored objects and colorants are colored, and why colors sometimes change when objects are viewed under different lighting conditions.

If the artists of today were to allow themselves to step, however hesitantly, from their own artistic world into the world of chemistry, they might find that they actually had the best of both worlds. They need not be concerned with manufacturing their own materials, nor with developing new materials for experimentation. They need not concern themselves with the quality of commercially available materials. The properties of these materials are well-known and available in the literature of artists' materials. An understanding of the chemical reactions that take place in the actual use of their materials is also available. All of this knowledge is often helpful in devising controlled artistic experiments in color with many of the haphazard, "hit-or-miss" elements removed. Such knowledge is helpful in simply having more control over the materials being used.

The purpose of this book is to open the world of the science of artists' colored materials to the artist. This will be done first by an examination of the nature of light and color for, without light, one cannot perceive artistic representation. Next, the nature of the matter that interacts with light will be examined, and then how light is modified by the objects that interact with it to produce color. Next, starting with some basic chemical principles, the nature and properties of organic and inorganic colored materials (colorants) will be discussed. The final part of the book consists of a series of sections on practical applications including the states of matter, solvents and solutions, artists' pigments, paints, dyes, fibers, polymers, ceramics, glasses, glazes and photography. A brief chapter on art hazards and how to avoid them concludes the book.

1.1 Selected Readings

Color Uncovered: An Interactive Book for the iPad
http://www.exploratorium.edu/downloads/coloruncovered/
If you've ever wondered what color a whisper might be, this delightful interactive book is for you. Created by the folks at the Exploratorium in San Francisco, "Color Uncovered" is a unique volume complete with articles, illusions, and videos that explore the art, physics, and psychology of color. Also, the book has some color activities that just require an iPad and basic items such as a drop of water and a piece of paper. This book is compatible with all iPads running iOS 4.3 and newer. [KMG]

CHAPTER 2

THE NATURE OF COLOR

Color is a property of materials that has been an integral part of human experience in every age and civilization. It has caused humanity to wonder about its origin and experiment in its production. Historically, the use of color was chiefly an art which developed slowly into an organized body of knowledge so that by the Fifth Century, B.C.E., the Greeks were writing treatises on color harmony, perspective and the preparation of pigments. They also succeeded in expanding the artist's palette to include white lead, red lead and vermilion. It was left to the practical, aggressive Roman businessman to commercialize color usage by the manufacture and distribution of "mass-produced" colored items. This was the beginning of a more advanced, but strictly empirical, color technology. However, it was not until Isaac Newton's (1642-1727) experiments in the Seventeenth Century that a firm theoretical foundation regarding the nature of color was laid.

Today, color science plays an important role in business, science and industry. It is one of the few disciplines that cuts across the boundaries of art, biology, physics, psychology, chemistry, geology, mineralogy, and many other fields. There is hardly an object or a substance in nature that is not colored, and virtually every commercially marketed item today is either deliberately colored or de-colorized.

Color is also the indispensable tool of the artist whether painter, decorator, fabric designer, ceramist or photographer. There is even a profession of color technologists whose task it is to identify and match colors for industry. Many industries are deeply concerned with color including the dyes used for tinting dry cereals, the wire coatings for the complex inner workings of a telephone routing or fiber optic system, and the screened dots of the color printers and television manufacturers. In industry, it is often necessary to match precisely huge batches of colorants, failing which a large amount of colorant may have to be discarded or turned to other uses.

To describe a color in words without the use of a color-name is impossible. When we call a color "red," most people know what we are describing because we have used a color-name within their immediate experience. However, if we attempt to describe it in any other way such as calling it light, warm, dark, cold, pure, dull, vivid, and so on, it could be any of an immense variety of red hues with varying degrees of contamination with black or white. We also know from experience that some colored objects seem to exhibit two different hues when illuminated with two different light sources, and even that different kinds of detection systems are more sensitive to some colors than to others. This is evidenced by the fact that cats are reputed to be able to "see in the dark" and that butterflies can respond to certain flowers that appear to be near-colorless to the human eye. Put another way, cats and butterflies have a wider response range to colors than human beings do.

From this discussion, it should be evident that it is important, when investigating the nature of color, to devise a scientific way to describe it and to measure it. We have already noted that color depends not only upon the colored object itself, but also upon the light in which it is viewed and the sensitivity of the viewer. Any complete description of color must take these three factors into account, a task that cannot be completed in this relatively short textbook. However, let us begin by examining the first experiments on color and by looking at some of the properties of visible light.

2.1 How Scientists Work

What is color? We cannot hold color in our hands. We cannot toss it from one surface to another. We cannot capture a handful of it from a rainbow, nor can we squeeze it into a hard mass. Experientially, we know that color is not a "substance" in the same way that wood and stone are substances, and yet it can inhere in these substances and change their appearance. A scientific probe of the nature of color involves examining how color changes the appearance of objects, and how color itself can be changed by its interaction with objects.

When scientists try to find out about nature, they always try to introduce some change by using an appropriate probe. Their primary purpose is to predict, which essentially involves forecasting the nature of a change. Information which does not involve change is not science; it is merely note-taking. Scientists are not interested in the static world.

The process of introducing change into a system, observing it, and carefully recording the information is called experimentation. The element of creativity enters into the selection of the change to be introduced. Often, the change is first observed by accident and is then deliberately reproduced in an experiment. This is what happened when Harvey Firestone (1868-1938) first mixed sulfur with natural rubber by error, only to discover that the product was no longer a gooey mass, but a resilient material that retained its shape. He followed up his observation with a series of experiments on the effect of sulfur on rubber, and succeeded in developing the first commercially viable rubber product.

2.2 Newton's First Experiment

Experiments with color, as with all scientific discoveries, are based on some prior observation followed by deliberate manipulation of the conditions that produced the phenomenon. However, how does one do this with as intangible a natural phenomenon as color? Theories regarding the origin of color go back to the ancient Greeks, and Aristotle himself is credited with making the first important contribution to what is now the modern theory of selective absorption. However, it remained for Isaac Newton in the Seventeenth Century to formulate modern color theory on the basis of experiment. In fact, Newton's first scientific publication was on the subject of light and color, after which he went on to discover the laws of motion, the calculus, and the law of gravity.

"A New Theory of Light and Colours," published in 1672, recounted Newton's experiments with prisms and light. Newton experimented with the prism during an enforced holiday due to plague while he was a student at Cambridge. Prisms were common objects in those days. Newton purchased his at a country fair. Although the properties of prisms had been observed many times before, it took Newton's uncommon mind to systematically uncover the secret of colored light by working with this simple, common object.

Newton darkened a room and made a small hole in the covering over a sunny window. The hole permitted a narrow beam of sunlight to pass through a suitably placed prism. The angle of approach of the sunbeam could easily be seen in the darkened room and could be measured. There, on the wall opposite the window, Newton saw the full glory of a shimmering spectrum, a sight that never fails to elicit admiration and wonder even today. The spectrum was a bright display made up of bands of colors which gradually changed hue from red to orange, yellow, green, blue and violet. A schematic diagram of Newton's experiment is shown in Figure 2.1.

Newton also saw what he called an "indigo" color between the violet and the blue regions of the color array, but no one else has ever seen it. The problem of color-matching, you see, is as old as the science of color itself.

Up to this point, Newton had done nothing more than had others before his time. It took seven more years and many more experiments before Newton felt that he was ready to publish his paper on a theory of color.

2.3 Newton's *"Experimentum Crucis"*

Let us review some of these experiments. It seemed from Newton's first experiment, reported and illustrated above, that sunlight was made up of a series of seven different colors (including indigo). Newton asked if these colors, in turn, were

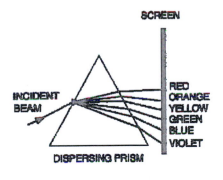

FIGURE 2.1

made up of mixtures of other colors. To find out, he allowed his spectrum to fall upon a screen with a vertical slit in it. Then he turned his prism until only red light passed through the slit. In back of the screen, he placed a second prism and allowed the red light to pass through it. Newton reasoned that if the red was a mixture of other colors, the second prism would disperse, or fan out, these colors in the same way the first prism had dispersed the white light into its constituent colors. No such dispersion into additional colors occurred, nor were any of the other colors dispersed into other colors when they were tried in turn. Each color appeared to be the same although, as expected, each was dispersed (appeared to have a wider band) to a greater degree by the second prism. A summary of this important experiment is illustrated in Figure 2.2. Newton concluded that ROYGBIV (the initials of the seven colors) were the fundamental colors recognized by the human eye and were not themselves mixtures of other colors. These so-called "spectral colors" are usually given in order starting with red, perhaps because red light is bent least by a prism.

Newton called this two-prism experiment his *"Experimentum Crucis,"* his crucial experiment. By it, he demonstrated conclusively that sunlight is a mixture of seven colors (including indigo), that these colors could not be further dispersed into additional colors, and must therefore constitute the fundamental colors.

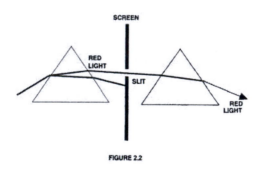

FIGURE 2.2

2.4 Newton's Recombination Experiment

In another of his experiments, Newton passed light through a prism and then allowed the dispersed beam to pass through a second prism with the reverse orientation (Figure 2.3). Given the fact that we expect the second prism in Figure 2.2 to further disperse the dispersed light, what do you suppose happens to the light beam in Figure 2.3 as it emerges from the second prism?

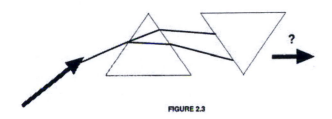

FIGURE 2.3

In this chapter, we have seen that white light, of which the most common example is sunlight, is made up of a mixture of colored lights which can be separated into an array of colors we call a spectrum. The fundamental colors into which sunlight is dispersed are called the spectral colors.

Miniexperiment I

Try generating a spectrum yourself. The best prism for this purpose is a 60° prism, but any faceted glass object may be used. Educational Innovations (www.teachersource.com; 888-912-7474), sells a variety of prisms. Lamp stores sometimes have pendants from crystal chandeliers available. Some inexpensive glass ashtrays have beveled edges that disperse light well. Thick plate glass obtained from waste pieces often has a fine, but exceedingly sharp, angled edge with a plane face. If all this fails, you can even use a Martini glass filled with water.

A large prism can be constructed from three glass plates of the same size glued with aquarium glue at the face edges to form a standing triangle. A fourth glass plate can be glued on as a base. When this is filled with water, it acts as a fine prism. By substituting other liquids for water, the ability of various liquids to disperse a beam of light can be compared.

A beam of light from a multimedia projector serves as a good source of light. You may wish to construct a 2" X 2" piece of cardboard with a narrow slit in the center to cover over the light beam of the projector. This serves to "collimate" the light into nearly parallel rays, a condition under which dispersion can best be observed. Alternatively, a powerful flashlight can be covered by a cardboard cone with the outermost tip cut away to leave a small, round hole. This method would most approximate Newton's original experiment.

Miniexperiment II

You may wish to repeat the experiment in Newton's *Opticks* (reference on page 16) which demonstrated that blue light is bent more than red by a prism (Fig. 2.1). To do this, take a piece of cardboard about 12" long and 4" wide. Glue a 6" X 4" piece of red construction paper to the top half and a similar piece of blue construction paper to the bottom half. Now wind a strand of black thread around each half in a wide spiral. Newton used the thread to make sharp lines on which to focus. You will also need an inexpensive magnifying glass. The lens in a magnifying glass has the same effect on a beam of light as the two prisms in Fig. 2.3. Now illuminate the red-blue surface with a projector lamp (Newton used a candle). Allow the light reflected from the red-blue surface to pass through the magnifying glass and to fall upon the surface of a piece of white cardboard. Maneuver the magnifying glass and the white cardboard until you have a sharp image of the thread on the red or blue side of the illuminated surface. If the thread on the red side is in focus, you will find that the thread on the blue side is out of focus because blue light is bent to a different degree than red light. You will have to move the white cardboard further away from the magnifying glass to focus the thread on the blue side, but then the red side will be out of focus. See Figures A, B and C on the following page. To replicate the conditions under which Newton worked, try using a candle.

16

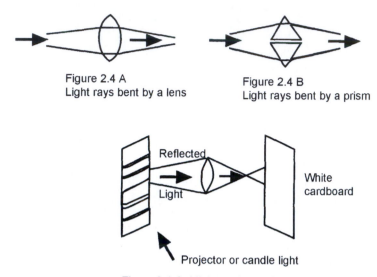

Figure 2.4 A
Light rays bent by a lens

Figure 2.4 B
Light rays bent by a prism

Reflected
Light

White
cardboard

Projector or candle light

Figure 2.4 C. Miniexperiment II

2.5 Selected Readings

Newton, I. (1704) *Opticks: or a treatise of the reflexions, refractions, inflexions and colours of light. Also two treatises of the species and magnitude of curvilinear figures.* Royal Society, London. Reprinted in Hutchins, R.M., Ed. Great Books of the Western World, Vol. 34; Encyclopedia Britannica: New York, 1952; pp. 386-412.

Orna, M.V. Chemistry and Artists' Colors. Part I. Light and Color. *J. Chem. Educ.* **1980**, *57*, 256-258.

Zollinger, H. *Color: A Multidisciplinary Approach*; Wiley-VCH: New York, 1999.

CHAPTER 3
VISIBLE LIGHT

3.1 What Is Light?

Before we can examine the fundamental nature of color more closely, we must ask about the nature of light since we saw in Chapter 2 that color can arise from white light. If you think about the properties of light for a moment, you may conclude that all light is accompanied by heat. You learn in childhood that glowing objects are hot. Numerous other examples of the heat-light relationship come to mind: the warmth of sunlight, the heat radiated by a light bulb, the red glow of electric heater coils and hot charcoal. We also know, at least at this moment, that the "cold light" of science fiction stories is not a present reality.

Since heat and light are so closely related and often arise one from the other, we may suspect that they share the same nature. That nature is called "energy," a familiar term and a familiar concept, but a term very difficult to define. For our purposes, we shall define energy as anything capable of causing a change in motion. Energy is required to move a stationary object, or to speed up or slow down an already moving object. Anything that causes these changes, or can be converted into something else that causes these changes, is a form of energy. You know that heat can be used to operate steam engines which, in turn, can drive other types of engine. You know that sunlight can be focused by a magnifying glass to cause intense heat, and eventually, burning.

3.2 Is Light a Particle or a Wave?

Energy occurs in many forms, and light is only one of these forms. Energy may also take the form of heat, electrical energy, mechanical energy, chemical energy or nuclear energy. Each of these forms can be converted into the others and certain devices in everyday life perform these conversions conveniently. An automobile battery converts chemical energy into electrical energy. A motor in a household appliance converts electrical energy into heat and light. By rubbing your hands together, you can convert mechanical energy into heat energy.

Each of these conversions requires some kind of interaction with material. Electricity requires a conductor of electricity. Chemical energy arises from the chemical reactions of the various chemical entities we call atoms and molecules. However, although light can arise from all of these interactions, it does not require a material substance for its transport. We know this because light from the sun reaches us by traveling though (nearly) empty space. Does this mean that light is different from the other forms of energy?

This question caused much controversy several centuries ago. Newton postulated that light is made up of little corpuscles or particles since it travels in straight lines, casts sharp shadows and changes speed as it passes through different forms of matter. On the other hand, his theory could not explain why light behaves in a markedly different manner from a stream of particles as it passes through a series of slits in material barriers.

Toward the end of the Seventeenth Century, a Dutch physicist named Christiaan Huygens (1629-1695) was experimenting with the properties of waves by examining the wave patterns of rippling water. He observed that the patterns caused by these waves traveling through the openings, or slits, in the barriers he placed in his ripple tank were identical to the patterns made by light when it was allowed to travel though similar sets of slits. This caused him to propose, in 1678, that light is a form of energy which has a wavelike nature, and that the wave characteristics of light can explain many of the properties of light that could not be explained by Newton's corpuscular theory. However, Huygens's wave theory could not explain how light manages to travel through empty space without the help of a supporting medium. Indeed, if one accepts Huygens's theory, one must also postulate that empty space is not empty, but is filled with some kind of invisible material, "ether," and that light is propagated through empty space on a series of ether waves.

3.3 The Wave Nature of Light

What is light made of? Huygens could not answer this question, but he could certainly explain many of the properties of light by his wave theory. Newton felt he could answer this question, but could not explain many of the observed properties of light by recourse to his theory alone. The wave-corpuscle dilemma had to wait until 1905 when two German physicists, Max Planck (1858-1947) and Albert Einstein (1879-1955), were able to show that light has a dual nature and that it behaves as particles in some ways and as waves in others, thus eliminating the need for "ether."

Using the ideas of Einstein and Planck, we may think of light as a special form of energy called electromagnetic radiation because a beam of light has been found to generate electric and magnetic fields[1] as it travels through space. Furthermore, as a beam of

---

[1] A field is a nonmaterial region in space which influences other regions and is capable of exerting a force. A magnetic field exerts a force on magnetic objects; an electric field exerts a force on objects with an electrical charge. Material objects which are nonmagnetic are transparent to magnetic forces. For example, the force of a magnetic field may be felt by an iron-containing object even when separated from the magnet by a piece of paper or cardboard.

light travels, the amplitude[2] of the field it generates at a fixed point has been found to oscillate in a periodic manner in much the same way that a pendulum oscillates back and forth about a central point. However, since a light beam is not stationary, but travels at a certain velocity, depending upon the medium, the oscillations may be viewed as being "spread out" into a series of waves. The vibratory motion of these beams can be likened in many ways to water waves. If you drop a pebble into a shallow pan filled with water, the disturbance the pebble creates will travel outward from the point of disturbance in a series of ripples we call waves. A wave is an up-and-down oscillation spread out over a distance.

To understand color more fully, we will have to examine the wave nature of light more closely. A beam of light of one color is in many ways like a series of waves traveling along a rope. If you hold one end of a rope, with the other end firmly attached to a wall, and move your hand vigorously up and down, you will observe wave motion along the rope (Figure 3.1). If you move your hand up and down more rapidly to create faster vibrations, you will generate new waves which are shorter in length that those you generated before. Figure 3.2 depicts the generation of a series of waves, where each series has successively shorter distances between the wave crests. If you now move your hand from side to side instead of up and down, the waves formed will be horizontal rather than vertical. In a similar manner, waves can be generated at a slant, always in a plane perpendicular to the direction the waves appear to be moving.

---

[2] Amplitude of vibration is defined as the maximum displacement from the undisturbed position. It can be determined by measuring the height of a wave peak from midpoint to crest (top of the peak).

Figure 3.1

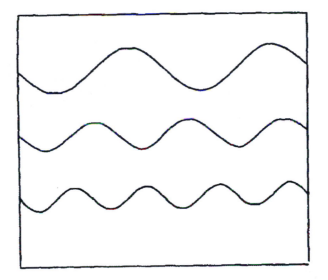

Figure 3.2

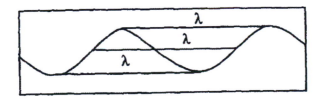

Figure 3.3

Light waves share many of the characteristics of rope waves. The waves can vary in length (wavelength) and in height perpendicular to the direction of wave motion (amplitude). However, light has no medium such as the rope; the electromagnetic disturbances can move through empty space. Also, the waves are not all vertical, or horizontal, or at any fixed angle in-between, but oscillate in all of these directions at one time.

A single wave of light may be depicted as in Figure 3.3. All of the horizontal lines in this diagram represent one **wavelength,** symbolized by the Greek letter λ (lambda). Thus a wavelength can be defined as the linear distance from a fixed point on the wave to an identical fixed point on the next wave. Wavelength is measured in units of length, such as centimeters (cm). However, since the wavelengths of visible light are so much smaller than one cm, it is more convenient to use a smaller unit of length. Several smaller units of length will be defined in the next section.

Notice that as the waves in a rope move forward along the rope, any given part of the rope itself moves up and down, but never forward. Each time a full wave progresses past a fixed point in the rope, that point moves up from a central position, reaches a maximum, moves back down through the central position to a bottom maximum, or minimum, and then moves back to the center again. The number of such oscillations measured in a unit of time, the second, is called the **frequency** of the wave. Your hand, which is supplying the energy, determines the frequency. The more energy you put into making the rope move, that is, the faster your hand vibrates, the greater the frequency. Frequency is measured in number of oscillations, or cycles, per second and is symbolized by sec^{-1} (read "reciprocal seconds." Another term for this unit is "hertz," where 1 hertz = one cycle per second (cps)).

Since the frequencies of visible light are of the order of 100,000,000,000,000 sec^{-1}, we will have to learn how to express such large numbers more conveniently in the next section.

Observe in our ropewave experiment that as the rope vibrates more rapidly, the length of each wave becomes smaller. In other words, as the frequency increases, the wavelength decreases. Frequency and wavelength are said to be inversely related. This is always true, regardless of the type of wave being examined. In addition, if the wave is traveling at constant velocity, the product of the wavelength and the frequency will determine the velocity. In words, this relationship is

$$\text{Wavelength times Frequency equals Velocity} \qquad [3.1]$$

If we symbolize wavelength by lambda, λ, and frequency by another Greek letter, v (nu), and velocity by the letter c, the relationship becomes

$$\lambda \; X \; v = c \qquad\qquad [3.2]$$

Finally, if we take wavelength in cm and frequency in sec^{-1}, we get cm X sec^{-1} = cm-sec^{-1} = cm/sec, where the unit of velocity is clearly distance divided by time. From this relationship, depending upon the material through which a wave is passing, which determines its velocity, we can always determine wavelength if the frequency is known, or the frequency if the wavelength is known.

3.4 The Frequency of Light Determines Its Color

Now that we know that a wave of light possesses both wavelength and frequency, let us see if we can measure these quantities. If we were clever enough to devise a scientific "ruler" capable of separating out and measuring each wave of visible light, we would find that different colors of light had different wavelengths. For example, a selected wave of red light has a wavelength of 0.00007 cm, but a particular wave of light that we perceive as "green" may have a wavelength of 0.00005 cm. These wavelengths are so small that about 19,000 wavelengths of visible light will fit across your thumbnail! If we continue our measurements on the wavelengths of various waves of visible light, we will discover several more amazing things:

1) Each wave of red light can have wavelengths that range between 0.0000647 and 0.0000700 cm, each wave of orange light has wavelengths that range between 0.0000585 and 0.0000647 cm, and so forth through the visible spectrum.

2) If we allow light to travel through different media, such as air, water, ethyl alcohol, *etc.,* we find that the wavelength and the velocity of a given wave changes from medium to medium, but that the frequency and the color remain unchanged.

From these facts, we conclude that each perceived color of visible light consists of a range of frequencies, and that color can now be described in terms of measured frequencies rather than in terms of some vague color name. However, since frequency and wavelength are related by equation 3.2, and since each color wavelength is invariant in the particular medium the light happens to be traveling through, a specification of wavelength may also be used to describe a wave of light when the medium is also specified. Since most research has been done in air or vacuum, media in which the velocity of light is almost identical, it has been customary to describe light waves in terms of wavelength.

3.5 A Digression on Exponential Notation

You have noticed that the wavelengths of visible light are quite small and that four zeros must be written after the decimal point and before the number that specifies the wavelength in centimeters. This is quite inconvenient. It would be better to be able to express wavelength in centimeters more concisely, or even express it in a different unit entirely.

Red light has a typical wavelength of 0.0000666 cm. If I were to divide the number 0.000666 by 10, the quotient would be 0.0000666. Similarly, division of 6.66 by 10 X 10 X 10 X 10 X 10 would also produce 0.0000666. Since 10 multiplied by itself five times can be expressed as $(10)^5$ or 10^5, then $6.66/10^5 = 0.0000666$. If you recall the rules governing exponents from your high school math days, you may remember that $1/10^5$ is the same as 10^{-5}. Thus 0.0000666 cm can be expressed as $6.66/10^5$ cm or as 6.66×10^{-5} cm. The latter quantity is the conventional way in which scientists express this small number. This method of notation allows us to keep track of all those zeros after the decimal point without actually writing them all down (and risking losing a few along the way).

Just as very small numbers can be expressed conveniently by using negative powers of 10, so very large numbers can be expressed by using positive powers of 10. A typical frequency for red light is 454,200,000,000,000 sec^{-1}. Multiplication of 4.542 by 10 multiplied by itself 14 times, $i.e.,$ 10^{14}, produces the same number. Thus, the rather cumbersome number given above can be expressed as 4.542×10^{14} sec^{-1}.

Sometimes, instead of changing the method of notation, it is more convenient to express a number in different units. For example, while it is convenient to use centimeters to measure the length of a sheet of note paper, this unit is certainly an inconvenient one when it comes to measuring distances between cities. It is also a very inconvenient unit for measuring very small distances like the wavelengths of visible light.

The fundamental unit of length in the metric system is the meter (m). All other units of length are either multiples or fractions by tenths of the meter. For example, a centimeter is 1/100 of a meter; therefore, there are 100 cm in 1 m. Similarly, a millimeter (mm) is 1/1000 of a m; thus, there are 1000 mm in 1 m. A mm is also 1/10 the length of a cm, so there are 10 mm in 1 cm. The wavelengths of visible light are so small that it is convenient to measure them not in terms of cm nor even in terms of mm but in terms of an even smaller unit that is only one billionth of a meter! This unit is known as the nanometer (nm); there are one billion $(1,000,000,000 = 10^9)$ nm in one m, or in 100 cm. It is now possible to convert 0.0000666 cm to nm using this relationship:

0.0000666 cm X (1,000,000,000 nm)/100 cm = 666 nm

Notice that all those zeros disappear and we have a handy number, 666, instead of 6.66×10^{-5} or 0.0000666. From now on, we will express wavelengths of light in nanometers, nm.

Another frequently used unit of wavelength is the Ångstrom (Å), a unit preferred by physicists. Ten Å make up one nm; therefore 666 nm = 6660 Å. If you wish to convert from Å to nm, just divide by 10.

3.6 The Properties of Light

Up to this point, we have seen that light has frequencies that can be measured in sec^{-1} and wavelengths that can be measured in nm. These two quantities are inversely related to one another and the product yields the velocity of the light (Equations 3.1 and 3.2).

We also saw in our ropewave experiment that frequency of vibration or oscillation is also related to energy. We saw that as we put more energy into making the rope oscillate, the rate of oscillation increased. This relationship is also true of light waves. As the frequency of light increases, its energy also increases. This is known as a direct relationship, mathematically, and the equation governing it was worked out by Albert Einstein and Max Planck, and is thus called the Einstein-Planck relationship:

$$E = h\nu \tag{3.3}$$

In this equation, energy is symbolized by E, ν is the frequency in sec^{-1} and h is a quantity known as the Planck constant, which has units of energy X time. A convenient value for h is 4.136×10^{-15} eV-sec, where eV is the symbol for a unit of energy known as the electron-volt. An electron-volt (eV) is defined as the energy an electron (to be defined later!) gains or loses when it moves through a potential of one volt. If, for example, each electron stored in an ordinary 12-V automobile battery has a potential energy of 12 eV, then this amount of energy is expended by each electron as the battery discharges in use.

Each color in the visible spectrum has associated with it a range of frequencies from which can be calculated corresponding wavelength ranges. According to Equation 3.3, it is now possible to calculate the energy associated with each frequency of light as well. The fundamental colors of visible light along with their properties (calculated from Equations 3.2 and 3.3) can best be summarized in tabular form (Table 3.1).

This table contains a wealth of information and it would be well for us to examine it in detail. First of all, we notice that just a very small change in wavelength effects a change in perceived color. Secondly, each color does not consist of a single wavelength, but of a wavelength range. Although the visible spectrum covers the 400 to 700 nm wavelength range, all of the rays from 400 to 424 nm are seen as violet, all of the waves from

TABLE 3.1

THE FUNDAMENTAL COLORS OF THE VISIBLE SPECTRUM†

Color	Wavelength (cm)	Wavelength (nm)	Band-Width (nm)	Frequency (sec^{-1})*	Energy (eV)
Red	0.0000647-0.0000700	647.0-700.0	53	4.634-4.283	1.92-1.77
Orange	0.0000585-0.0000647	585.0-647.0	62	5.125-4.634	2.12-1.92
Yellow	0.0000575-0.0000585	575.0-585.0	10	5.214-5.125	2.16-2.12
Green	0.0000491-0.0000575	491.0-575.0	84	6.103-5.214	2.53-2.16
Blue	0.0000424-0.0000491	424.0-491.0	67	7.071-6.103	2.93-2.53
Violet	0.0000400-0.0000424	400.0-424.0	24	7.495-7.071	3.10-2.93

*These values must be multiplied by 10^{14}

†Since the dividing line between perceived spectral colors is difficult to discern, their subdivision into six broad regions is somewhat arbitrary. Listed below is another division suggested by Arthur C. Hardy in Handbook of Colorimetry; Cambridge: The MIT Press, 1934; p. 2.

Red	700-610 nm	Green	570-500 nm
Orange	610-590 nm	Blue	500-450 nm
Yellow	590-570 nm	Violet	450-400 nm

491 to 575 nm are seen as green, and so forth. These individual ranges are called pure spectral colors. It is difficult for the eye to distinguish between waves with wavelengths of 495 nm and 560 nm. Both waves *are* perceived as green. The wavelength range accounts for the "smear" effect that you see when white light is dispersed by a prism.

A third feature that we notice from the table is that some colors have broader wavelength ranges than others. For example, the wavelength range of green light is 8.4 times greater than that of yellow light. This fact is also evident when white light is dispersed by a prism. However, if you measure the widths of the green and yellow bands with a ruler, you will find that the green band is not 8.4 times the width of the yellow band. This is because a prism does not disperse each wavelength of light to the same extent. The term "band-width" is used to denote the wavelength range of a given color.

Another important feature gleaned from this table is the energy values of visible light. The last column gives the energies associated with each color of light. We see that red light has an energy range of 1.77 to 1.92 eV, whereas violet light has an energy range of 2.93 to 3.10 eV. Violet light is considerably more energetic than red light. This means that violet light packs a larger wallop with respect to any substance with which it interacts.

You can see by examining the table that the energy differences between colors are quite small, in fact, exceedingly small because electron-volts themselves are very small units. This shows what a very sensitive instrument your eye is. To distinguish between red and orange requires a remarkable ability to discriminate between very small energy differences, yet it is a sensitivity we routinely take for granted. Did you know that the human eye can distinguish between five and eight million different colors?

3.7 The Velocity of Light

All colors of light and all light waves travel through empty space at the same velocity. You probably learned in your grammar school days that this velocity is 186,000 miles per second. This is an inconceivably large velocity, amounting to 5,865,696,000 miles traveled by light in one year. This is the so-called "light year," used by astronomers to measure the distances between heavenly bodies. The light year is actually a unit of distance, not of time. It points to the fact that although light travels quickly, it still takes a finite amount of time to move from point to point. For example, if you look up at the star Sirius tonight, you will see the star only because you see the light that has emerged from it. Since Sirius is four light years away from earth, this means that the light you see tonight left Sirius four years ago. It has been traveling through space four years, and is just now reaching your eye. This means, too, that you will see Sirius tonight as it was four years ago. In the meantime, it is possible that, say two years ago, Sirius could have exploded, but you will not know about that event until two years have elapsed because the light recording the explosion is still two light years away. In other words, because light moves with a finite velocity, it takes time for it to travel. Because of the great distances associated with outer space, when you look up at the heavens and gaze at the stars tonight, you are literally gazing into the past.

However, our present concern is with the velocity of light and its measurement. Since light travels at such a tremendous velocity, you can imagine how difficult it is to measure. Galileo

Galilei (1564-1642) tried back in the Sixteenth Century and failed. He thought he could apply the same method to the measurement of the velocity of light that he had used successfully to measure the velocity of sound. For the latter measurement, he and a colleague stood several hundred yards apart. Galileo fired a gun and immediately began to note the time it took for him to hear the answering shot fired by his assistant, who was instructed to fire as soon as he heard Galileo's shot. Galileo, noting the exact time he heard the second shot, reasoned that the total time elapsed between firing the first shot and hearing the second was the time it took for the sound to travel the known distance to his assistant and back again. A simple calculation gave Galileo a very satisfactory result for the velocity of sound, which is about 750 miles per hour at room temperature.

Galileo next thought that he could use the same method to measure the velocity of light, but this time by uncovering a bright lantern at an exactly known time. When he tried this experiment, he found that the time of uncovering the light and the time of sighting it were virtually one and the same. This meant, of course, that light traveled much too fast over the distance in question to be measured by ordinary means. It remained to the ingenuity and sophistication of the American physicist, Albert Abraham Michelson (1852-1931), in our own timeframe, to make the first satisfactory measurement of the velocity of light. As you know, this turned out to be 186,000 miles per second, or in the metric system, 3.0×10^{10} (thirty billion) cm/sec or, more exactly, 2.997925×10^{10} cm/sec. This figure actually represents the velocity of light traveling in a vacuum. As we shall see in the next chapter, the change in the velocities of the different wavelengths of light as they enter a different medium accounts for the prismatic light spectrum with which we are already familiar.

It is no wonder that Galileo failed in his attempt to measure this velocity because light travels so fast that it can circle the earth more than seven times in one second. Development of methods for the precise measurement of this quantity was still very active research even into the late 1950s. Furthermore, Galileo lacked the very accurate timepieces with which we are now familiar and was forced to measure very short time periods by singing! Michelson himself must have sung a song or two when he learned that he had won the 1907 Nobel Prize in Physics for his measurements of this fundamental quantity.

3.8 Selected Readings

Bragg, Sir William *The Universe of Light;* Dover Publications: New York, 1959; Chapter 1

Drake, S. The Role of Music in Galileo's Experiments. *Scientific American* **1975**, *232*(6), 98.

The Nobel Prize in Physics 1907
(http://www.nobelprize.org/nobel_prizes/physics/laureates/1907/michelson-bio.html)

Miniexperiment III

The relationship between light colors and energies may be demonstrated with an incandescent bulb and a light dimmer or voltage regulator. In a darkened room (no stray light), start with the dimmer or regulator dial at zero and gradually increase it. What color light appears first? What appears next? (You will not see green light because the green light adds itself to the red to give more yellow; see Section 10.6). If you have the use of a hand spectroscope, repeat the experiment but, this time, observe the portions of the spectrum which successively appear.

Chapter 4

The Bending of Light

Both the artisan and the
experimental scientist share a
manipulative heritage that often
makes them cousins, if not brothers
and sisters, at the workbench.

John B. Eklund
"Art Opens the Way for Science,"
Chemical and Engineering News,
June 6, 1978, p. 32

CHAPTER 4
THE BENDING OF LIGHT

Miniexperiment IV

Add water to any transparent container with straight sides such as a baking dish or a square or rectangular jar. A fish tank is excellent for this experiment, even if it is occupied. Insert a pencil partway into the water, holding it vertically. Observe it at eye level while slowly moving it horizontally to the right or left. You will observe that the part of the pencil above the water line appears to be separated from the part that is immersed in the water. This phenomenon is due to refraction, or the bending of light.

Miniexperiment V

Cover a flashlight with a cardboard cone with a small opening at the tip. This will create a small pencil of light. Now, all you need is a fishtank with slightly dirty water (the fish won't interfere with the experiment) and a darkened room. Make the air dusty by lighting a match and blowing it out to develop smoke near the surface of the water, or by gently clapping a dusty board eraser above the water. Direct a pencil of light upward to the middle of the tank from the lower right or left side. By changing the angle of incidence, you will see light bent at various angles, and even bent totally back into the water. Also, try directing the beam from the top to the middle bottom of the tank. Lower the beam by changing its angle while directing it to the same spot. Observe the path of the light beam as it passes the interface between air and water.

4.1 Refraction

The bending of light as it passes from one medium such as air to another such as water, glass, oil, *etc.,* is called refraction. Refraction occurs when light travels from one medium to another. This is because the velocity of light is affected by the medium through which it passes. Light slows down as it passes from air to other media, and speeds up when it travels from other media into air.

We shall see shortly that the phenomenon of refraction can help us understand how a prism separates white light into its constituent colors. By having recourse to the laws of refraction, we will also be able to better understand other observations such as the apparent breaking of the pencil in Miniexperiment IV, or why your leg appears shorter as you step into water. Can you think of any work of art which exhibits refraction?

The reason why objects are visible to us is that light striking them is reflected to cause an image of these objects on the retina of the eye. The fish in the fish tank of Figure 4.1 is visible to the observer for this reason. As the reflected light, which travels in straight lines in a given medium, reaches the water-air interface, it is bent in the direction depicted in the figure. Since the observer perceives this bent light ray as if it were traveling in a straight line, the fish is sighted along this straight line and appears to be nearer to the surface than it really is. Native American spear fishermen knew this through experience and always aimed their spears at a more acute angle than seemed necessary.

To understand why a beam of light is bent as it travels from one medium to another, we will use the analogy of a moving car. The car represents the light beam, and the ground on which it travels represents the medium. As long as both front wheels of the car travel on paved ground, it can move ahead traveling in a straight line. However, if one wheel is slowed by loose sand, the other wheel still on the pavement moves faster, and the car turns in the direction of the sand. Once both wheels are in the sand, the car can then proceed with both wheels turning at the same rate

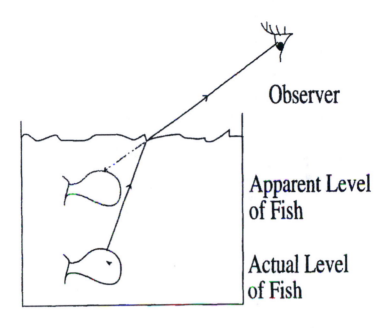

FIGURE 4.1

again. It will move more slowly through the sand, but it will be able to maintain a straight line in a new direction. If both wheels encounter the sand simultaneously, the car slows down, but does not change direction.

A beam of light, composed of a group of parallel waves, is analogous to the moving car. All of the waves in the beam travel at the same velocity in a straight line when moving through a given medium such as air. If the beam strikes the surface of the water at an angle, the waves in the beam that strike the water first will be slowed down, while the waves that have not yet struck the surface will continue to travel at their original velocity. The waves already in the water are analogous to the automobile wheel spinning in sand; the others are analogous to the wheel still on the hard road surface. Consequently, the light beam will be bent in the direction of the water. Once the entire wavefront of the beam has entered the water, the beam will continue in a straight line, but in a new direction. The amount of bending is directly related to the amount of slowing down of the light in the medium. If the velocity of light in a given medium is less than that in air or vacuum, the medium is said to be more "optically dense" than air or vacuum. Some media are more optically dense than others. Diamond is the most optically dense material known.

We have seen that when light travels from a less dense to a more dense medium, the beam bends in the direction of the more dense medium. If the direction of travel is reversed, the light beam, as it emerges from a more dense into a less dense medium, will also reverse its behavior and bend away from the less dense medium. Verify this behavior by mentally reversing the direction of travel of the automobile in the analogy we just used.

You can see a similar. direction-velocity relationship if you watch a chorus of dancers (such as the Rockettes at Radio City Music Hall in New York City) on a stage. To cause the line of dancers to turn, the dancer at one end steps in place while the others progressively down the line move faster and faster so the whole line rotates.

4.2 The Prism Effect

The bending of light by a glass prism can be explained in terms of refraction. Picture a beam of light entering a prism at an angle (see Figure 4.2). Since glass is more optically dense than air, the light entering it slows down and the beam bends into the glass and away from the direction in which it was originally traveling. In terms of the automobile analogy, the right wheel is slowed down. In order to describe what is happening more precisely, let us draw an imaginary surface at right angles, i.e., perpendicular, to the surface of the prism. This is called the "normal" to the plane and is represented in Figure 4.2 by the dotted line, CJ. As the incident beam of light, AB, enters the prism, it is bent towards the normal and travels in a new direction, BF. Now, when the refracted beam, BF, emerges from the prism at point F, it is entering a less dense medium and will therefore reverse its behavior. If it bent toward the normal, CJ, when entering the prism, it will now bend away from the normal, GE, upon leaving. Thus, its new direction is FH once it has passed beyond the prism.

The two optical surfaces of our prism in Figure 4.2 are not parallel to one another. For this reason, the beam of light, as it emerges from the prism, is bent even further from its initial direction than it was when it entered the prism.

When a light beam interacts with an object with parallel optical surfaces, such as a piece of plate glass, the beam follows the same laws of refraction, but the outcome is quite different, as can be seen in Figure 4.3. When the light beam enters the glass pane, it is bent toward the normal, as shown. When it emerges, it is bent away from the normal, but since both normal planes (dotted lines) are parallel to one another, the emergent beam ends up traveling in a path parallel to its original path. The result is that one bending of

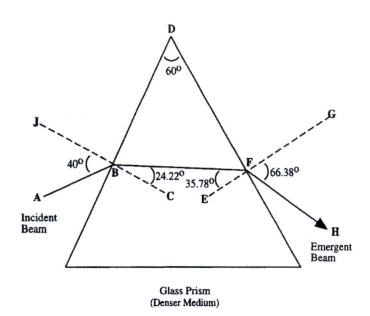

FIGURE 4.2

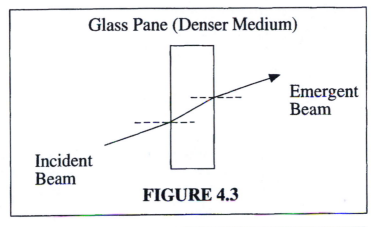

Glass Pane (Denser Medium)

Emergent Beam

Incident Beam

FIGURE 4.3

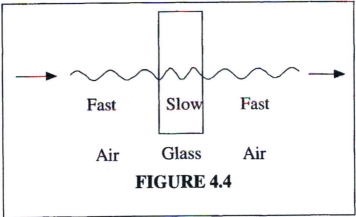

Fast   Slow   Fast

Air   Glass   Air

FIGURE 4.4

the light beam cancels the other. We will see in Section 4.4 that the two different directions of light beams, one emerging from the prism, and one emerging from the plate of glass, will help us to understand why a prism can disperse visible light and a piece of plate glass cannot.

4.3 Refractive Index

The greater the change in velocity as light enters a new medium, the greater the degree of bending, or refraction. A quantity which is defined in terms of these phenomena is the refractive index, which can be represented in two ways:

1)   Refractive Index = $\dfrac{\text{Velocity of Light in Air}}{\text{Velocity of Light in Medium}}$ [4.1]

The number in the numerator of equation 4.1 will always be larger than that in the denominator because the velocity of light is greater in air (or vacuum) than in any other medium. Thus, the refractive index will always be greater than one.

2)   Refractive Index = $\dfrac{\text{Sine of Angle of Incidence}}{\text{Sine of Angle of Refraction}}$ [4.2]

Equation 4.2 needs more explanation than equation 4.1. The sine of an angle, if the angle is in a right triangle, is the quotient of the length of the side opposite the angle and the length of the hypotenuse. In Figure 4.2, for example, we can create a right triangle by connecting points A and J. The angle ABJ is the angle that the incident beam makes with the normal and is defined as the angle of incidence. The sine of the angle of incidence equals the length JA divided by the length AB. This value would go into the numerator of equation 4.2. You can, instead, find the sine of the angle ABJ in a Table of Sines or by using your pocket calculator if it has trigonometric functions. Suppose the angle of incidence, ABJ, were 40°. Using a Table of Sines or entering 40 and hitting the "SIN" button on your calculator, you see that the sine of 40° is 0.643, so let us put that number into the numerator of equation 4.2.

Now, the angle of refraction is defined as that angle that the refracted beam makes with the normal. From Figure 4.2, we can see that this angle is CBF. Measurement of this angle will yield a value somewhat smaller than that of the angle of incidence if the material is more dense than air. Let us suppose that this measured angle is 24.2°. The sine of 24.2° is 0.41, and this value goes into the denominator of equation 4.2. A quick division yields a value of 1.6 for the refractive index of our prism. A glance at Table 4.1, which gives the refractive index values for several media, indicates that our prism is probably made of light flint glass. We could have used equation 4.1 to calculate the refractive index of our prism as well. We know that the velocity of light in air is 3.0×10^{10} cm/sec, so that value can be substituted into the numerator. Now, all we have to do is to measure the velocity of light as it travels through the prism and substitute our measured value into the denominator. You will recall from the discussion in Chapter 3 that the velocity of light is not an easy quantity to measure but, using some kind of clever device, let us say that we come up with a value of 1.9×10^{10} cm/sec for our prism. Substitution of this value into the denominator of equation 4.1 followed by division yields a value of 1.6 for the refractive index. Both equations yield the same results. Equation 4.2 is more difficult to understand, but measurement of angles of light bending is quite easy. However, once the refractive index of a medium is known via equation 4.2, then calculation of the velocity of light in that medium via equation 4.1 is very simple. As the refractive index of a medium increases, the velocity of light in that medium decreases. Diamond, with a refractive index of 2.47 (Table 4.1), has the highest refractive index known. Hence, light experiences its greatest slowing down when it enters a diamond. Because diamond has such a high refractive index, it is possible to cut a diamond in such a way that all of the light entering the face facets of the diamond are reflected back again to the viewer. This is why properly cut diamonds are so brilliant; they are capable of totally reflecting the light striking them. (Keep in mind that some of the light striking the diamond is also absorbed. The totality of reflection refers to that portion of the light which would normally be transmitted by the diamond.)

Equation 4.2 also tells us that refractive index is related to the degree to which light is bent when it enters a medium. The greater the refractive index of the medium, the greater the degree of bending of the light as it enters that medium.

4.4 Light Bending and Wavelength

When light slows down by entering a medium more dense than air, its frequency remains unchanged, but its wavelength becomes shorter. This effect can be compared to someone swimming in water as compared to, let's say, swimming in thick maple syrup. For the same number of strokes and the same energy, the swimmer will cover much less distance in the maple syrup. This change in wavelength is illustrated in Figure 4.4.

4.5 Refractive Index and Frequency

The degree to which a medium affects the velocity of light depends not only upon the density of the medium, but also upon the frequency of the light. If light of several frequencies (several colors) enters a medium, the light of higher frequency is slowed down to a greater degree and therefore experiences a greater degree of bending. Blue light, with a higher frequency range than red light, is bent to a greater degree by both a prism and by a piece of plate glass. However, as the light emerges from the plate glass, all the frequencies of light are bent back again with the blue once again bent more than the red. Hence, they will emerge traveling in a path parallel to the incident pathway; all the frequencies will end up traveling parallel to one another and dispersion will not take place.

TABLE 4.1

AVERAGE REFRACTIVE INDEX OF SELECTED MEDIA AT 20 °C FOR SUNLIGHT

Medium	Refractive Index
Vacuum	1.000
Air	1.0003
Water	1.333
Alcohol (Ethyl)	1.360
Linseed Oil	1.49
Light Flint Glass	1.6
Diamond	2.47

As red and blue light emerge from a prism, however, the geometry of the prism brings about an even greater bending of the two frequency ranges. Since the two colors emerge from the prism at different angles, they can be viewed separated from one another. Hence, dispersion takes place. Table 4.2 gives the average velocities and refractive indices for ordinary glass for each color of visible light.

As we mentioned previously, a very interesting example of how knowledge of the laws of refraction can be applied is in the cutting of precious stones and semi-precious gems. Most gemstones exhibit the brilliance they do because of the high refractive indices of the substances of which they are composed. Flint glass has a refractive index of about 1.6 for white light, whereas diamond has a refractive index of 2.47 and zircon about 1.9. Because diamond's refractive index is so large, light rays exhibit such small angles of

TABLE 4.2
REFRACTIVE INDICES FOR THE FUNDAMENTAL
COLORS IN LIGHT FLINT GLASS

Color	Velocity of Light (cm/sec)	Refractive Index
Red	1.925×10^{10}	1.567
Orange	1.910×10^{10}	1.571
Yellow	1.905×10^{10}	1.575
Green	1.893×10^{10}	1.585
Blue	1.882×10^{10}	1.594
Violet	1.859×10^{10}	1.614

refraction in diamond that the rays will often fail to leave the material, but will be reflected back into it. This phenomenon is known as total reflection, and it can be achieved in a diamond if the stone is properly cut. When a diamond exhibits total reflection, none of the light entering the face of the stone will be able to escape out the back, but will be totally reflected back to the face, imparting a brilliance and sparkle to diamond that has only been matched recently by synthetic gems of almost equally high refractive index.

4.6 The Rainbow

One of nature's most breathtaking displays of the spectrum is the rainbow. A rainbow can be seen because finely dispersed water droplets are suspended where they can act as tiny prisms and disperse the rays of sunlight. You can see a rainbow if you send a fine spray from a garden hose into the air against a dark background when the sun is behind you. The size of the droplets affects the color range that you see.

In a later chapter, we shall see how the refractive index is an important characteristic in determining the opacity of a pigment.

Refraction produces many other light phenomena such as mirages, the apparent "steam" or "heat" escaping from a radiator surface on a cold winter's day, the illusory appearance of wet pavement on a hot, dry road, the "threads" that develop when mixing alcohol and water, and many other sometimes bizarre sights.

4.7 Selected Readings

Bragg, Sir William *The Universe of Light*; Dover Publications: New York, 1959; pp. 80-84.

Davis, Jefferson C., Jr. Introduction to Spectroscopy, Part III. *Chemistry* **1975**, *48* (May), 19-22.

Minnaert, M.*The Nature of Light and Color in the Open Air*. Dover Publications: New York, 1954; p. 238.

Chapter 5

The Electromagnetic Spectrum

"Isn't it astonishing that all
these secrets have been preserved
for so many years just so that we
could discover them?"

Orville Wright, in a letter of June 7, 1903,
six months and ten days before
the first powered flight by a human being

CHAPTER 5

THE ELECTROMAGNETIC SPECTRUM

5.1 Electric and Magnetic Fields

In section 3.3, we asked the question, "What is light made of?" The answer to such a question can only be arrived at by examining the properties of light, *i.e.,* by observing how it interacts with its surroundings. Careful observation reveals that each energy packet of light, as it oscillates through space, generates an electric field and a magnetic field at right angles to one another. We are all familiar with the idea of a magnetic field. We know, for example, that our earth has a magnetic field with terminals at the magnetic north and south poles. Small magnets can be purchased as toys and large magnets find many uses in science and industry. The chief characteristic of a magnet is the fact that it possesses an invisible "field" of force around it which enables it to exert its influence on another magnet over a distance. As an example, if you slowly move a small magnet toward an iron-containing metal object, such as a paper clip, at some distance away from the magnet the clip will suddenly react to the force field of the magnet by actually leaping across the intervening space and sticking to the magnet.

Electric fields are just as familiar to us from everyday experience. Synthetic fabrics have a tendency to "cling" under certain conditions; woolen blankets tend to snap, crackle and pop when we fold them after laundering.

If you perform Miniexperiment VI, you will find that you can create your own electric force field and that this field has properties similar to those surrounding a magnetic pole. Magnets have directional characteristics terminating in "poles," and these poles have been dubbed "north" and "south." The north pole of one magnet is repelled by the north pole of a second magnet, but is attracted by the south pole. This observation

---

Miniexperiment VI

Cut out two discs of aluminum foil about 1/2 inch, or 1 cm, in diameter. Suspend each disc about 12 inches apart from one another by thread about 12 inches long. This can be done by wedging the loose ends of the thread between or under books on a shelf.

To generate an electric charge, rub a piece of fur or plastic wrap on a glass or plastic rod. Bring the charged rod carefully between the two discs. Do the discs move? In what direction?

Re-charge the rod and touch it to one of the discs. Now, re-charge the rod again and carefully bring it close to the same disc. What do you notice this time?

Before the rod touched the discs, it exerted an attractive force on them. After contact, the rod has shared its charge with the disc. Both have the same charge and should repel one another.

Try charging the rod to see if it will attract bits of tissue paper, grains of salt or sugar, or cornflakes. What determines whether objects will be attracted upwards to the rod?

Run a very thin stream of water from a faucet. Hold the charged rod alongside the stream. What do you observe?

---

leads us to the general conclusion, "opposites attract, likes repel." Similarly, there are two types of force fields among electric fields, and these have been dubbed "positive" and "negative" since they are opposites of one another. With respect to electric fields, properly called electrostatic fields since the electric charges are not in motion, a positively charged object attracts a negatively charged object, and *vice versa*.

Light is electromagnetic radiation. As such, it generates an electric field and a magnetic field. Since these fields are of the same nature as the magnetic and electric fields with which we are more familiar, they should also be able to interact with fields in material objects and indeed they do. However, since the electromagnetic field of light oscillates with a certain frequency characteristic of a particular energy packet of light, we might expect that the frequency of oscillation will dictate how light and matter interact.

Most of you are familiar with an object called a tuning fork. Tuning forks can be manufactured to emit a certain note when struck, for example, the tuning fork that gives the same tone as "middle C" on the piano. No matter how often or how hard you strike the "middle C" tuning fork, it will always give out the same note. This is because this tuning fork vibrates at a certain frequency, 256 oscillations per second, and this is called its "natural frequency of vibration." Two objects that have the same natural frequency of vibration are capable of inducing vibration in one another, even over a distance. When a piano tuner adjusts the middle C string of a piano, he/she can tell that it is adjusted perfectly when the vibrations from a middle C tuning fork cause the string to vibrate without striking the piano key. Hence, we see that sound oscillations from one object can affect another. Similarly, light waves from one object can affect another. For example, we know that long exposure to sunlight causes fading in some dyes and pigments. This means that the pigment and dye molecules must have been changed by the action of light. The very fact that light is "visible" must mean that it is causing some kind of disturbance on the retina of the eye; otherwise, we would not even know that it was around.

5.2 The Electromagnetic Spectrum

When we observe some of the ways in which light interacts with matter, we find that matter interacts with certain sources of light differently from its interaction with others. For example, I can be exposed to fluorescent light all day long and not experience sunburn. If I wish to get a good tan, I must expose myself to sunlight or a sun lamp. Both sources are visible light, but sunlight seems to contain a component capable of interacting with matter (my skin!) more vigorously than fluorescent light can. Is it possible that sunlight might contain additional frequencies of electromagnetic radiation that are not necessarily visible but still share the same characteristics of visible light?

To test this idea, let us devise a simple experiment. We all know that photographic film is light-sensitive. If we allow visible light to fall upon a piece of photographic film, it becomes exposed, and development will produce a photographic image. Let us pass sunlight through a prism so that the several colors are dispersed and, in a darkened room, let us place a piece of photographic film just beyond the violet portion of the spectrum. Development of the film will show that it was exposed even though the agent causing the exposure was invisible. This agent, radiation beyond the violet end of the spectrum, is called ultraviolet radiation. Though it is undetectable by our eyes, its presence can be detected by devices that are sensitive to high frequency electromagnetic radiation. The ultraviolet region was discovered by Johann Wilhelm Ritter (1776-1810) by using a light-sensitive silver compound.

In 1800, an astronomer named Sir William Herschel (1738-1822) performed a similar experiment at the opposite end of the spectrum. He was interested in determining the heating effects of the various colors so he dispersed sunlight with a prism and placed thermometers in each of the spectral regions, one in the red, one in the orange, etc. He also placed thermometers beyond the red and violet regions and discovered, much to his amazement, that the highest temperature was recorded for the region that lay beyond the red region of the spectrum. Thus, Herschel is credited with having discovered what is now called the infrared region. Note that the prefix "infra" means beneath, or below. Infrared rays have frequencies and energies which are less than those of visible radiation, hence "infra." Ultraviolet rays have frequencies which surpass those of the violet portion of the visible spectrum and are called "ultra," which means exceeding, or beyond the moderate. We see that although the human eye is a very sensitive probe of reality, it has limits and those limits are, as implied in Chapter 3, approximately 400 and 700 nm, respectively; the human eye is blind to electromagnetic radiation that lies beyond these wavelengths.

The entire range of frequencies of electromagnetic radiation is called, collectively, the electromagnetic spectrum. For convenience, it is divided into regions which are named either with reference to the visible spectrum (infrared, ultraviolet), with reference to their source (X-rays, gamma rays, cosmic rays), or with reference to their use or frequency (radio waves, microwaves).

Electromagnetic radiation can interact with matter in much the same way that a tuning fork interacts with a piano string. The fundamental units of most substances, called molecules, all have certain natural frequencies of vibration. If electromagnetic radiation with similar frequencies interacts with these molecules, the radiation, although invisible to our eyes, can induce vibrations in the molecules. Other, lower frequencies of electromagnetic radiation can cause the molecules to rotate end over end; certain higher frequencies of radiation can cause such severe disturbances in the electric field of a molecule that the molecule can literally "fall apart."

Can the electromagnetic radiation we call light create all these disturbances? We know from experience that light can interact with matter to create at least some of them, and we shall devote much of this text to examining the way light interacts with matter to produce the visible color spectrum.

5.3 Range of the Electromagnetic Spectrum

Figure 5.1 is a schematic diagram of the frequency range of electromagnetic radiation showing the relationship of the visible region to the entire range.

Using the frequency range of Figure 5.1, we can calculate that the visible region comprises less than 0.00005% of the range of electromagnetic radiation. In other words, the human eye is an instrument sensitive to only 0.00005% of all the waves that could possibly impinge upon it, and it is these very few frequencies to which we assign the term "light" in the strictest sense of the word. We can now define visible light as that region of the electro-magnetic spectrum to which the human eye is sensitive. The other regions of the spectrum are invisible though each has its own use and its own detection devices, thanks to modem technology. Those in the radio range, for example, are of rather low energy and are used to broadcast radio and television programs. You know that WCBS New York is "Newsradio 88." This means that the station broadcasts at 88 kilocycles per second, or $88,000$ sec^{-1}, a frequency well inside the radio region. Waves in the far, middle and near (to the visible) regions are used by scientists to identify chemical compounds, as are the rays in the ultraviolet and far ultraviolet. Since X-rays and gamma rays are so highly energetic (50 to $1,034,000$ times more energetic than visible radiation), they are

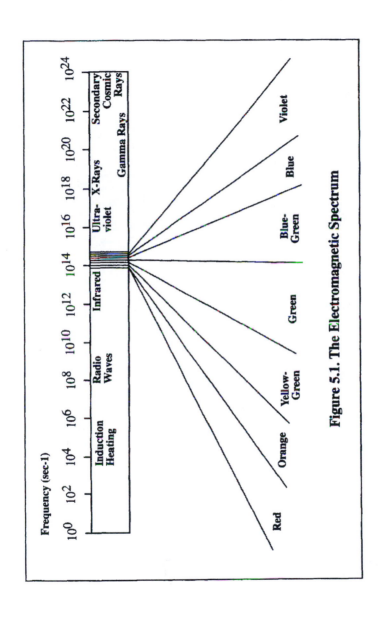

Figure 5.1. The Electromagnetic Spectrum

capable of penetrating normally opaque materials such as brick walls and human flesh. For this reason, they can be used for medical purposes such as X-radiographs and radiation treatments for malignant tissues. The divisions between the regions are not clear-cut, however. A commonly accepted division of the regions with respect to wavelength, frequency and energy is given in Table 5.1.

To get an idea of the scale of the electromagnetic spectrum in everyday foot-yard units, let us use a scheme proposed by Edward Bowen (see Section 5.4). If we multiply our wavelength scale by three million, 100 nm become one foot, and atoms become as large as periods on a page. The optical microscope would focus particles between two miles and one yard across. Red, green and blue light would have wavelengths of 7, 5 and 4 feet, respectively, while ultraviolet rays would have a wavelength range of between 2 and 4 feet. Infrared radiation would have wavelengths between about 25 and 250 yards.

Hal Hellman (see Section 5.4) has compared electromagnetic waves and sound waves as follows: A piano keyboard has its notes arranged to run from A to G, and then to start all over again. Each series from A to G is called an octave. Each letter in an octave has twice the frequency of the same letter in the octave below. The electromagnetic spectrum also has its series of notes but, whereas a piano has seven octaves, the electromagnetic spectrum has some seventy "octaves." Visible light is only one small "octave" in the range. The frequency of violet light at one end of the visible range is almost twice that of the red light at the other end. Yet, this one "octave" of light holds the secret of life without which chlorophyll could not absorb the energy needed by green plants. It also holds all of the wavelengths that yield the immense variety of color.

TABLE 5.1
THE ELECTROMAGNETIC SPECTRUM

Region	Wavelength Range (nm)	Frequency (sec-1)	Energy (eV)
Radio Frequency and Microwave	22,000,000	1.364×10^{11}	0.00056
Far Infrared	15,400 – 22,000,000	$1.364 \times 10^{11} - 1.948 \times 10^{13}$	0.00056-0.081
Mid Infrared	2,500 – 15,400	$1.948 \times 10^{13} - 1.200 \times 10^{14}$	0.081-0.496
Near Infrared	700 – 2,500	$1.200 \times 10^{14} - 4.283 \times 10^{14}$	0.496-1.77
Visible	400-700	$4.283 \times 10^{14} - 7.495 \times 10^{14}$	1.77-3.10
Ultraviolet	200-400	$7.495 \times 10^{14} - 1.500 \times 10^{15}$	3.10-6.20
Far Ultraviolet	10-200	$1.500 \times 10^{15} - 3.000 \times 10^{16}$	6.20-124.1
X-Ray	0.01-10	$3.000 \times 10^{16} - 3.000 \times 10^{19}$	124.1-124,000*
Gamma Ray	0.0005-0.140	$2.143 \times 10^{18} - 6.000 \times 10^{20}$	8,900-2,500,000*
Secondary Cosmic Rays†	0.0005	6×10^{20}	2,500,000

* X-rays and gamma rays are distinguished from one another on the basis of their origin, the former arising as the form of energy when electrons in atoms "spin off" energy, and the other when the heavy center of the atom, the nucleus, de-excites. The two regions overlap somewhat.

† Cosmic "rays" are actually high energy particles from outer space which interact with the earth's atmosphere to produce electromagnetic radiation called secondary cosmic rays.

5.4 Selected Readings

Bowen, Edward J. The *Chemical Aspects of Light*; London: Oxford at the Clarendon Press, 1946; pp 43-44.

Frercks, J.; Weber, H.; Wiesenfeldt, G. Reception and Discovery: The Nature of Johann Wilhelm Ritter's Invisible Rays. *Studies in History and Philosophy of Science* **2009**, *40*,143-156.

Hellman, Hal *The Art and Science of Color;* McGraw-Hill Book Company: New York, 1967.

Herschel, W. Investigation of the Powers of the Prismatic Colours to Heat and Illuminate Objects. *Phil Trans Roy Soc London* **1800**, *90*, 255-283.

Rainwater, Clarence *Light and Color;* Golden Press: New York, 1971.

CHAPTER 6

WHITE LIGHT

6.1 The Spectrum of Sunlight

Although a knowledge of the fundamental nature of light and color is indispensable to all those who use color, it is very little help when trying to produce, describe and reproduce the color we actually *see*. To do this, we have to know much more about the nature of the light sources we use, the interaction between the light source and the colored object, and the nature of the detection system. We must take into account not only the physical production of color, but also the psychophysical reaction to it. We must deal not only with pure colors in an abstract situation, but with mixed colors and the results of color mixing in substances. We must distinguish between colored light and colored objects; we must quantify and measure. In a word, we must objectify our examination of color with a view to gaining control over the conditions whereby color is produced and modified. This entire endeavor comes under the heading of the vast field of color technology, portions of which will be treated in more detail in Part II.

One step toward gaining control over color production and modification is an understanding of the light source being used. So far, we have virtually confined our discussion to the source of light we term sunlight. Frequently, in this book, we have referred to sunlight as "white light," although the term is open to question. We know that the quality of sunlight is different at different times of the day and under different atmospheric conditions. We also know that sunlight viewed from the moon is quite different from sunlight viewed through our atmosphere. So "white light," or "sunlight," is very difficult to define.

Since the light we see is composed of all the visible wavelengths, it is possible to describe that light by measuring the relative intensity or energy emitted by the source at each wavelength. For example, with the proper instrument, I can make a series of measurements of the energy coming from typical sunlight at appropriate wavelength intervals and record the results in a table. Now, although this table (Table 6.1, on the following page) describes typical daylight almost exactly, it is very difficult to visualize what this distribution looks like when presented in tabular form. Scientists generally like to "draw" the relationship between two variables such as intensity and wavelength by plotting such data as a graph. If we let the Relative Intensity be the vertical, or y, axis, and the wavelength in nm be the horizontal, or x, axis, we can plot each point of our table as shown in Figure 6.1. If we then connect the points by drawing the best line to get a smooth curve, we get the plot shown in Figure 6.2. This plot is called a spectral energy distribution curve; it gives a "profile" of the energy distribution of the light source in question. By examining Figure 6.2, we can see that typical sunlight exhibits its greatest intensity in the blue region of the spectrum, and that the intensities fall off gradually as we look toward the infrared and ultraviolet regions. From our discussion of Herschel's discovery and the darkening of photographic film by the ultraviolet component of sunlight in the previous chapter, we know that this is true.

6.2 Blackbody Radiation. One very important source of radiation from the scientist's point of view is the so-called "blackbody," which is defined as a heated source whose spectral

TABLE 6.1

SPECTRAL ENERGY DISTRIBUTION OF TYPICAL SUNLIGHT

Wavelength (nm)	Relative Intensity*	Wavelength (nm)	Relative Intensity
400.00	20.7	554.29	25.7
408.57	22.0	562.86	24.9
417.14	23.2	571.43	24.4
425.71	23.0	580.00	23.8
434.29	22.4	588.57	22.7
442.86	26.4	597.14	22.4
451.43	28.9	605.71	22.4
460.00	29.5	614.29	22.3
468.57	28.9	622.86	21.5
477.14	28.9	631.43	21.0
485.71	28.3	640.00	21.0
494.29	27.7	648.57	20.1
502.86	25.3	657.14	20.2
511.43	26.8	665.71	20.7
520.00	26.5	674.29	20.3
528.57	26.7	682.86	18.3
537.14	26.4	691.43	17.8
545.71	26.1	700.00	18.4

*Light emitted by a source can be described in terms of relative intensity at each wavelength. The relative intensities in this table are adjusted to a value of 25.0 at 560 nm. Absolute intensities are measured in terms of amount of energy per unit time falling on a given area for each wavelength in the spectrum. A typical absolute unit would be watts per square centimeter per nanometer.

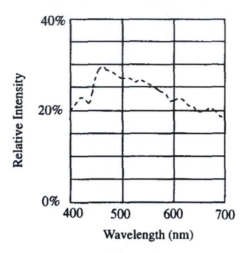

FIGURE 6.1

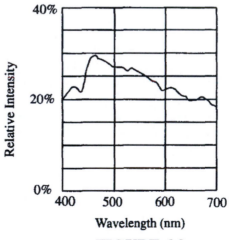

FIGURE 6.2

energy distribution curve (and therefore, color) is a function of its temperature alone. You are quite familiar with the fact that as objects, particularly metals, are heated up, you can gain some idea of the degree of hotness by simply observing the color of the object. An iron poker, for example, is black at room temperature. As it is heated in a fire, it radiates invisible infrared radiation and later begins to glow red. We all know that further heating can cause it to glow white. Similarly, the coolest stars are red, and as they increase in hotness, they undergo color changes from red to orange to yellow to white to blue. A red star typically has a temperature of around 2700 °C, whereas a blue star may be as hot as 40,000 °C. All other star colors lie somewhere between these two extremes of temperature. In all of these cases, we can measure temperature by measuring color via the spectral energy distribution curve of the glowing bodies themselves. This phenomenon allows us to correlate color with temperature so that, as a matter of fact, the temperature of a blackbody is called its color temperature. (See Miniexperiment III, Chapter 3).

6.3 Spectra of Other Light Sources

Unfortunately, most of our common light sources, including daylight and the various forms of artificial light, are not blackbodies. There are such wide variations in their spectral energy distribution curves that a knowledge of the curve of each source is necessary in order to be able to ascertain what color will be seen when the source interacts with certain objects.

An incandescent light bulb, perhaps one of our most familiar artificial light sources (although soon to be phased out because of energy considerations), consists of a tungsten wire enclosed in a bulb filled with an inert gas. When the wire has electrical energy passed through it, it becomes hot and begins to glow. Figure 6.3 displays the spectral energy distribution curve for a 100 watt General Electric tungsten lamp. You can see that the radiation is most intense in the red region and on into the infrared region, but that the intensity in the blue-violet region is very weak. The incandescent bulb is said to be rich in red light, but poor in blue-violet light. We perceive the light coming from this type of source as a warm reddish-yellow. The fact that an incandescent bulb emits far more energy in the infrared region than in the visible is the cause of its "downfall" – this heat energy is so inefficient and such a waste of energy that for environmental and practical reasons, this type of source will be discontinued.

A fluorescent tube, on the other hand, depends upon the principle of fluorescence rather than incandescence. The latter depends upon high temperature to produce the radiation, whereas fluorescence occurs when light of shorter wavelength is absorbed to produce light of longer wavelength. For example, ultraviolet light can be absorbed and then a small part of the absorbed energy can be successively lost in small bits (like a ball bouncing down steps) to become heat energy, after which the remainder is released (like the ball then falling out a window) in a burst of energy. The burst of emitted light has less energy than the light initially absorbed, so it has a longer wavelength and falls in the visible region of the spectrum. This conversion to light of longer wavelength is called fluorescence and takes place in less than one ten-thousandth of a second after irradiation. Because of the extremely short interval between irradiation and emission, fluorescence ceases when the irradiation ceases. A fluorescent tube is coated on the inside with a special powder made of a mixture of fluorescent solids called phosphors. A small amount of mercury is also placed inside the tube. When high voltage is applied, electrons in the mercury vapor are energized and emit this

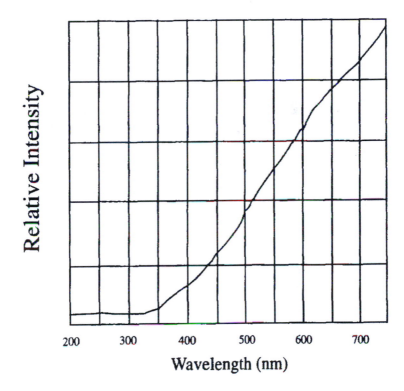

FIGURE 6.3. Spectral Energy Distribution
Curve for a 100 Watt G.E. Light Bulb

energy as ultraviolet light. The ultraviolet rays, which are invisible and also cannot pass through the glass, excite the phosphors which in turn emit visible light. By varying the phosphors, the white light produced can be varied as well.

Fluorescent tubes are more efficient than incandescent bulbs because much of the energy supplied to the electric bulb is wasted as heat, as pointed out earlier. For this reason, improved compact fluorescent lamps will gradually replace the inefficient incandescent bulbs.

The fluorescent tube does not heat up. A fluorescent lamp has a spectral energy distribution curve which shows some discrete (individual or distinct) lines superimposed upon a continuous spectrum (Figure 6.4).

A fluorescent tube called "Ultralume" is now gaining in popularity because it is even more economical than the usual tubes. The light given off is unique in that it consists of only three bands of color - red, green and blue - which appear as three sharp peaks on its spectral energy distribution curve.

There are many sources of light which emit only parts of the visible spectrum. When such a source is observed, only the colors that are actually emitted can be perceived. One such source is the sodium lamp; its spectral energy distribution curve is shown in Figure 6.5. Notice that there is a sharp spike of high relative intensity at around 589 nm, and several other spikes of somewhat lower intensities at around 340, 425, 505 and 575 nm. Since the intensity at 589 nm is so great, in comparison with the other wavelengths emitted, that is the wavelength that will be perceived by the eye. An examination of Table 3.1 indicates that this spike falls in the yellow-orange region of the spectrum, so the color perceived will be yellow-orange.

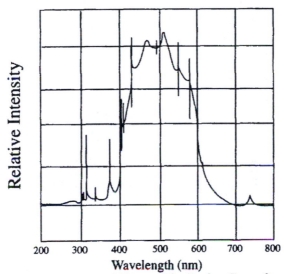

FIGURE 6.4. Spectral Energy Distribution Curve for a G.E.
15 Watt Standard Cool Fluorescent Lamp

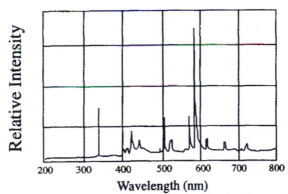

FIGURE 6.5. Spectral Energy Distribution Curve for a
Sodium Lamp

Figures 6.2, 6.3 and 6.4 depict the spectral energy distribution curves of three sources of so-called white light. Each of these curves is very different from the others, and yet in each case the eye perceives the light emitted from these sources as "white." However, since each of these sources is different, each will interact differently with colored objects, which accounts for the oft-observed phenomenon of colored objects exhibiting different shades of color when viewed under different light sources. This effect will be discussed in more detail in a subsequent chapter.

However, before we undertake this important task, we must now examine the nature of matter since it is matter that produces light and color in the first place.

6.4 Selected Readings

Berns, R.S. *Billmeyer and Saltzman's Principles of Color Technology*, 3rd ed. Wiley: New York, 2000.

Feller, R (1964) "Control of the Deteriorating Effects of Light upon Museum Objects," *Museum* , **1964**, *XVII* (2), 85-98.

Part II

Matter and Color

"The scientist does not study
nature because it is useful; he
studies it because he delights
in it, because it is beautiful."

Jules Henri Poincaré,
Foundations of Science,
New York: Science Press,
1920; p. 18

CHAPTER 7

INSIDE THE ATOM

Miniexperiment VII

Before beginning this chapter, which will be about the atom, you are going to perform what is called a "Black Box" experiment. In this experiment, a group of you will examine three sealed boxes and will be expected to describe the contents without opening them. Several other groups will examine other such sets of boxes with contents identical to those in the first set. You may be able to tell how many objects are in the box, what the shape of each object is, whether it is relatively hard or soft, and perhaps the size and material composition of each object. You may gently shake the box, turn it around, weigh it if you wish, or otherwise examine it with any of the tools in the room, but you must not damage the box or open it. After you have decided what is in the box, appoint one member of the group to record it on paper. This is the only miniexperiment that requires other persons besides yourself, and needs an instructor.

7.1 Matter and Energy

We saw, in Part I, that many objects, when excited by heat energy, electrical energy, or other forms of energy, give off light. In fact, we saw that all light originates in some material object whether that object be a glowing wire, a burning candle, excited phosphors, or the sun itself. Energy, in the form of light, cannot be divorced from matter. Therefore, it should help us to understand light and color better if we can understand the nature of matter better. We might be able to answer such questions as why objects emit different colors at different temperatures, why the color of a tungsten lamp is different from the color of a fluorescent lamp, or why a mercury arc lamp has a different color from a sodium arc lamp.

It may surprise you to learn that the effort to answer these questions, which you may have thought important only to the artist or color technologist, opened the door to all of our knowledge about the nature of matter. Locked in the spectrum of emitted light is a whole world of information about the submicroscopic world that we shall never see directly.

7.2 Higher Energy States

Heat and light are both forms of energy. When an object is heated to produce light (and more heat), the object does this by absorbing the heat energy and converting part of it into light. When a material object absorbs energy in this way, it is raised to a higher energy state. When it gives up the energy in the form of light, or other form of energy, it returns to a lower energy state.

The raising of the energy state of an object is analogous to a baseball that has just come into contact with a swung bat and enters the air above the bat as a high pop fly. The ball travels up as a result of energy imparted to it by the bat's momentum. At the instant the ball reaches the peak of its flight, it has attained its maximum energy of position (called potential energy). It has absorbed this energy by interaction with an energetic object, the bat. If there were some way for the ball to remain suspended in mid-air, it would retain this absorbed energy by virtue of its position. However, in real life, what goes up must come down. The ball must return to the ground, that is, the lowest possible energy state. In so doing, it must give up its newly acquired energy in the form of mechanical impact with the ground or with the fielder's mitt.

When a particle of matter interacts with energy, essentially the same phenomenon occurs. The radiant energy comes within the interaction field of the particles and imparts its energy to a particle which is then raised to a higher, or excited, state. After a longer or shorter period of time, the particle must return to the "ground" state, that is, to the lowest available energy state. In the process of this return, it must give up its excess energy in the form of radiant energy. There is only one important difference between this occurrence and the baseball analogy. When the baseball reached its maximum height, the amount of potential energy (energy of position) it possessed at that point depended entirely on how much energy had been transmitted to it by the bat's impact (neglecting losses though friction, *etc.*). If the impact had been very small, the ball would not have traveled very far. The ball can soak up varying amounts of energy depending upon how much is available; the ball is limited only by the amount of impinging energy. Not so with a particle as small as the atom! An atom is severely limited in the amount of energy it can absorb, and although this is a universal phenomenon, such limitation can only be observed on the atomic level.

This energy limitation, as we shall see shortly, is ultimately responsible for the colors in the light emitted by an excited particle. To understand the natural energy limitation imposed on particles as small as atoms, it is necessary to learn more about the nature of the atom. The remainder of this chapter will be devoted to examining the nature of the fundamental unit of all matter, the atom.

7.3 Fundamentals of Matter: Atoms and Elements

You probably already know that all matter in our universe is made up of combinations of only about 100 or so different kinds of atoms (as well as some breakdown parts of atoms). Even after many years of familiarity with this fact, it still seems miraculous that all the familiar and more exotic forms of matter that we encounter in our everyday lives - the leaves on trees, the spider in its web, brilliant tropical fish, human blood, tennis balls, cement, velvet, glue, glass, eggs, scissors, wood, oil, eyes - all are formed from various combinations of only a few types of atoms.

Each type of atom is called an element. Scientists have thus far discovered or synthesized (or believe they have synthesized) 118 different elements. The smallest particle of an element, that smallest unit that is identifiable as the element, is what we call the atom. This is analogous to saying that the smallest unit of humanity is the person. We may find people linked into couples, families, societies, nations, but we cannot deal with humanity in units smaller than one person. If a person were exploded into separate parts - legs, arms, eyes, etc. - then one would no longer be dealing with a person. In the same way, an atom is the smallest unit of an element in which an element can retain its identity. Chemical reactions deal with atoms and aggregates of atoms, and how they change partners. If an atom were attacked by an extraordinarily large amount of energy, or were inherently unstable, as are radioactive atoms, then the atom might split apart. However, we would no longer be dealing with the same element; the new particles formed by this type of change would have new identities.

Some examples of elements are carbon, gold, iron, nickel, copper, radium, barium and lead. The element hydrogen is composed of hydrogen atoms; the element carbon is composed of carbon atoms. Hydrogen atoms are fundamentally different from carbon atoms. If I were to take a single atom of carbon and try, by some means or other, to divide it up into simpler particles, I would no longer have carbon if I were successful in my endeavor. The atoms of any one element differ from those of other elements. How many elements, in addition to those already mentioned, can you name?

Although atoms were postulated as long ago as the Fifth Century, B.C.E., it was only in the last century that we came to regard their existence as fairly well proven. The problem is that there is no way that our normal detection devices can perceive atoms because they are extremely small. There are trillions of atoms in your fingernail alone. If you think about it, you will find it just as easy to picture an element such as iron as composed of continuous "iron" matter as to imagine it composed of tiny "iron" particles which somehow stick together. Yet today, although no one has ever seen an atom, the evidence that all matter is composed of atoms is thoroughly convincing, not only through logical conjecture based upon experimental evidence, but also through bits of pictures obtained by many different types of instruments, each of which reveals some singular restricted view of matter.

When John Dalton (1766-1844), in the early 1800's, first convinced the scientific world of the existence of atoms, it was believed that they were rigid little balls that had no simpler working parts. Then, about 100 years later, the striking discovery was made, largely through the use of the same type of device that now makes up our modern television tube, that atoms are made up of sub-units that are common to all different types of atoms.

7.4 The Parts of the Atom

To determine why an iodine atom is different from a copper atom, we must peek inside these atoms and look at their parts. Atoms, like everything else in nature, have parts which are arranged in the definite order we call structure. All atoms consist of three basic particles: protons, neutrons and electrons. Protons are extremely small compared to the atom itself. The region inhabited by the protons and neutrons, called the nucleus, is so small that if a typical atom could be enlarged to the size of a classroom, the nucleus would be the size of a period on this page. The remainder of the volume of the atom is composed of electrons.

Although electrons occupy so much space in the atom, they contribute almost nothing to its mass.* Each electron has a mass about 1/2000 as much as a proton (1/1837, to be more exact).

Since the proton has such a small mass in terms of grams, it is more convenient to devise a weight scale based upon the mass of one proton, since a proton is a fundamental unit of all atoms and contributes very much to the total mass. This new weight scale is the atomic weight scale, or more correctly, atomic mass scale and, on this scale, a proton is assigned a mass of 1.0078252 unified atomic mass units (abbreviated u); for our purpose, a mass of 1.0 u is close enough. Since it takes 1837 electrons to make up the mass of one proton, each electron must have a mass of 0.000544 u (verify this number). If we round this number off to only one decimal place, what conclusion do we draw regarding the contribution of electrons to the total mass of an atom?

Electrons repel one another with an electrostatic repulsion in much the same way that the like poles of two magnets repel one another, or that the charged rod repelled the aluminum disc in Miniexperiment VI. Likewise, protons repel one another, but they attract electrons. We conclude that the charges of electrons and protons are different since opposites attract and likes repel. Despite the vast difference in mass between the proton and electron, they each have the same quantity of charge, even though the charges act in opposition to each other. An electron attracts a proton with the same force that it repels another electron. The charge on the

---

*Mass refers to the amount of material substance present in a body. The standard unit of *mass* is the kilogram. 1 pound = 0.45359237 kilogram.

proton has been assigned a plus sign, +, while that of the electron has been given a negative sign, -, to distinguish between them. The total charge of the proton has been set at +1, so the charge on the electron, since it is equal in magnitude but opposite in sign or effect, is -1. When a positive charge and a negative charge of equal magnitude are brought together, they act together as if they have no charge at all. The net effect is that the charges cancel each other to produce zero charge.

The neutron, as you might guess from its name, behaves as if it had no charge. It is neither attracted nor repelled by protons or electrons. The neutron has an exact mass of 1.0086654 u, which also rounds off to 1.0 u. Thus, the neutron has about the same mass as the proton. The properties of the three fundamental subatomic particles are summarized in Table 7.1.

Next we shall consider how these particles are distributed in atoms to make the elements different from one another.

TABLE 7.1
THE FUNDAMENTAL PARTICLES

Particle	Mass	Charge
Proton	1	+1
Neutron	1	0
Electron	Negligible	-1

7.5 Subatomic Particles

Listed in Table 7.2 are data on some representative elements. Examination of this table reveals some interesting information. The first six elements listed are the six lightest known elements. Hydrogen, with one proton, is the lightest. Helium, the next lightest element, has two protons and two neutrons. Lithium follows with three protons and beryllium with four. If we keep adding protons and naming a new element for each additional proton, up to 92, we will have named all of the naturally occurring elements, the last of which is uranium. The remaining 26 elements (up to 118 protons) are synthetic. (Note: Elements with proton numbers of 43 and 61 are extremely rare in nature and were originally discovered by synthesizing them.)

The number of protons in the nucleus of an atom is called the atomic number. The atomic number determines the identity of an element. All atoms of lithium, for example, have three protons in their nuclei. All atoms of carbon have six protons, all atoms of gold have 79 protons, *etc.* Therefore, the element with atomic number 6 is carbon, and the element with atomic number 79 is gold, by definition. The element with atomic number 82 is lead. If I were able to pluck three protons from an atom of lead, I would have succeeded in changing lead into gold. Thus, modern chemistry points the way to fulfilling the dreams of the ancient alchemists (although nature spontaneously goes in the opposite direction).

A neutral or "complete" atom of each element has the same number of electrons as protons. This is because electrical neutrality demands an equal number of positive and negative charges, so in order for an atom to be neutral, it must contain the same number of protons and electrons. Hydrogen has one electron per atom, helium two, and so on. How many electrons are contained in one neutral atom of the element with atomic number 50?

All elements except hydrogen have neutrons in their nuclei. The neutrons apparently act as a kind of nuclear glue to hold the protons together in the tiny volume of the nucleus. As atoms get heavier, the number of neutrons increases more rapidly than the number of protons. Neutrons do such a good job of binding the nuclear components together that stable nuclei never come apart at the energies involved in chemical reactions.

Next, we see that the mass of each atom depends on the number of protons and neutrons present. The mass of the electrons, from Table 7.1, is negligible. If hydrogen has one proton in its nucleus, it has a total mass of 1 u. Helium, with two protons and two neutrons in its nucleus, has a total mass of 4 u. Thus, one

TABLE 7.2
SOME REPRESENTATIVE ELEMENTS

Element	Number of Protons	Number of Neutrons	Number of Electrons	Mass Number	Symbol
Hydrogen	1	0	1	1	H
Helium	2	2	2	4	He
Lithium	3	4	3	7	Li
Beryllium	4	5	4	9	Be
Boron	5	6	5	11	B
Carbon	6	6	6	12	C
Iron	26	30	26	56	Fe
Gold	79	118	79	197	Au
Lead	82	125	82	207	Pb
Deuterium	1	1	1	2	^{2}H
Tritium	1	2	1	3	^{3}H

helium atom is four times the mass of one hydrogen atom. You see that you can calculate the relative masses of atoms in u by simply counting the total number of protons and neutrons, a sum called the mass number.

Finally, each element has its own symbol, consisting of either one or two letters. The first letter is always capitalized and the second is always lower case when two letters are used. The symbols of most of the elements are derived directly from the first initials of their names or from two key letters if the initial is repeated. For example, the symbol for boron is B; the symbols for the other elements beginning with the letter "b" are Ba, Be and Bk for barium, beryllium and berkelium, respectively. The symbols for some of the elements known from ancient times are derived from their Latin names such as Cu for copper *(cuprum),* Pb for lead *(plumbum),* Au for gold *(aurum)* and Ag for silver *(argentum).* The symbols for several other elements are derived from their former names such as K for potassium *(kalium)* and W for tungsten (*wolfram*).

7.6 Isotopes

The last two entries of Table 7.2 are listed as deuterium and tritium. Each atom of deuterium has one proton and each atom of tritium also has one proton. Each of them has an atomic number of one. We know that hydrogen also has an atomic number of one, and that this number identifies all atoms of hydrogen. This means that deuterium and tritium must be forms of hydrogen. If we look at the third column of the table, we see that whereas hydrogen has no neutrons, deuterium has one neutron and tritium has two neutrons. Thus, the total atomic masses of hydrogen, deuterium and tritium are 1, 2, and 3. Each member of this trio has the same atomic number but a different atomic mass. Such groups of atoms are called isotopes. Isotopes are defined as atoms of the same element with the same atomic number but with different atomic masses. Hydrogen (sometimes called ordinary hydrogen to distinguish it from its isotopes), deuterium and tritium are all the same element.

Normal carbon, as listed in Table 7.2, has six protons and six neutrons in its nucleus. There is a much rarer form of carbon that has 8 neutrons and a total mass of 14 u. These two forms of carbon are symbolized as ^{12}C and ^{14}C. The three isotopes of hydrogen discussed above can be symbolized by ^{1}H, ^{2}H and ^{3}H.

There is a rare form of mercury, ^{197}Hg, that has the same mass as an atom of ordinary gold, ^{197}Au. Thus, two atoms of two different elements can have the same atomic mass; what differentiates them from one another is their different atomic numbers.

Isotopes of an element exhibit almost identical chemical behavior because chemical behavior is determined by the atomic number, that is, the number of protons in the nucleus and the number of electrons outside the nucleus in the neutral atom. Hydrogen, however, is an exception since addition of a neutron to ordinary hydrogen doubles its mass, and addition of two neutrons triples its mass. This tremendous relative mass change, in each case, causes the isotopes of hydrogen to differ very slightly in their chemical behavior. Thus, alone among the elements, the isotopes of hydrogen have special names to indicate their differences. Tritium, incidentally, is radioactive.

A quantity affected by the existence of isotopes is the relative atomic mass of each element. Although the relative atomic mass of the lead listed in Table 7.2 is 207 u, not all lead atoms have a *mass* of 207 u. A typical lead deposit in nature contains 1.48% ^{204}Pb, 23.6% ^{206}Pb, 22.62% ^{207}Pb and 52.3% ^{208}Pb. The weighted average of all of these isotopes amounts to 207.2 u. Since most elements have several naturally occurring isotopes, most of them have fractional atomic masses.

Now that we know something about isotopes, we can also go back to the reason why a proton has an assigned mass of 1.0078252 u and why we have been referring to a relative atomic mass scale. The question is "Relative to what?" We have been referring to the proton as having a mass of 1 u, and it would have been possible to set the mass of one proton to 1 u exactly, *i.e.*, 1 with an infinite number of zeros following the decimal place. Then, we could have weighed every atom relative to the proton. This would be one form of a relative atomic mass scale. However, chemists and physicists have agreed to use the isotope of carbon with a mass of 12 (^{12}C) as the international relative atomic mass standard. They set the mass of ^{12}C at 12.000000000000exactly u, and then weighed all other atoms relative to it. On this scale, a proton turns out to have a mass of 1.0078252 u, a neutron has a mass of 1.0086654 u, ^{9}Be has a mass of 9.012186 u, and so forth.

We remarked earlier that it is amazing that all of the matter in the universe is made up of only 92 naturally occurring elements. Now we see that all of the elements are made up of only three different kinds of particles. From the print on this page to the cosmic dust at the farthest reaches of the universe, ordinary matter is made up of only these three types of particles. Nuclear scientists are now trying to determine what these particles, in turn, are made of. This appears to be a much more complex mystery involving bits of matter and energy called quarks and charms. How this puzzle will be unraveled, if ever, remains to be seen. Scientists sometimes speak of the "onion" of knowledge, with layer hidden under layer. The more layers we peel away, the more infinitely large the onion appears to be.

7.7 Selected Readings

Leary, J.; Kippeny, T. (1999) A Framework for Presenting the Modern Atom. *Journal of Chemical Education* **1999**, *76,* 1217.

Orna, M.V. (1978) The Chemical Origins of Color. *Journal of Chemical Education* **1978**, *55,* 478-484.

Roundy, W. What is an element? *Journal of Chemical Education,* **1989**, *66,* 729-730.

CHAPTER 8

THE EXCITED ATOM

8.1 Line Spectra

Illuminants give off light when energy is supplied to them. However, when the source of energy is shut off, they stop glowing. Let us isolate a single element, say neon, supply it with enough energy to cause it to glow, and then pass the emitted light through a prism. What do we expect to see?

One way to energize elements is with electricity. If we pass a sufficient quantity of neon gas into an evacuated tube and then apply high voltage to the tube through electrodes sealed in the glass, a beam of light will appear in the tube. This beam of light, when first passed through a slit in a screen and then through a prism, will yield a continuous spectrum like that of the sun (Figure 6.2). At lower applied voltage, parts of the continuous spectrum fall away, leaving narrow, shimmering separated thin lines of light, a line spectrum. Every element has a characteristic line spectrum which is always the same for a given element and which differs from that of every other element.

There must be some connection between the unique line spectrum of an element and the interior arrangement of its atoms. In other words, the color display that can be used to identify an element must be related to the internal structure of its atoms.

8.2 Electronic Transitions Cause Visible Spectra

Until about 1900, the atom was considered to be a hard, little ball which was the smallest particle of matter. Only at the turn of the 20th century was it discovered that the atom is made up of sub-units. It was later discovered that one of these sub-units, the electron, was the particle that caused the colored line spectra observed for each element. The key to how electrons were arranged within atoms came directly from the study of the line spectrum of

---

Miniexperiment VIII

Set up a diffraction grating in one end of a cardboard tube with a narrow slit at the other end. Gratings, which are finely ruled pieces of glass or plastic, are preferable to prisms because of their high dispersing power. Consult any general physics text for an explanation of how a grating works; its mechanism of dispersion is quite different from that of a prism.

Spectra may be generated from an energized tube of gas or by heating salts of various metals in a Bunsen burner flame. (If a high-voltage gas tube is used, do not proceed without instructions from a supervisor.) Dissolve salts (either nitrates or chlorides) of potassium, sodium, copper, lithium, barium and calcium in a little water in a series of labeled beakers. Dip the tip of a nichrome wire (you can scavenge some from an old toaster – it works well for these flame tests because it does not melt or oxidize in a burner flame) in one of the solutions and then hold the wire in the outer part of the Bunsen burner flame. Note the emitted color. Next, look at the emitted light through the cardboard tube by peering at it through the end of the tube with the diffraction grating. Draw the line spectrum observed using colored pencils. Clean the nichrome wire by dipping it into some dilute hydrochloric acid and then into the flame. Repeat this procedure until the wire produces no emitted color. Then go on to examine the line spectrum of the next element.

hydrogen. The existence of these line spectra led eventually to the quantum theory of matter, a view of the mysterious ultimate nature of matter that is very different from our everyday experience of it. Color thus provided the key that opened the door to our modern ideas on matter.

What causes line spectra to appear? They appear when electrons in atoms, having first absorbed energy from an external source, then give off that energy. Each electron is constrained to absorb and emit only certain discrete quantities of energy. The electrons are promoted to higher energies (become excited) by absorption of heat energy, electrical energy or light energy from an external source. When an electron undergoes a transition back to a lower energy state, the difference in energy between the two states is emitted in the form of electromagnetic radiation which we, in some instances, perceive as color when the energies happen to fall in the visible region of the electromagnetic spectrum. (Electronic transitions can also take place in the ultraviolet but these are invisible to us.) When electrons de-excite from many different higher energy states, the lines merge to give a continuous spectrum.

By measuring the frequency of the light emitted, the energy differences between electron levels of higher and lower energy can be calculated. From these differences, a picture of the electron arrangements in the atom can be deduced. Each "packet" of energy that an electron absorbs is called a quantum. One of the cornerstones of the quantum theory is that energy can only be absorbed as individual or distinct "packets" called quanta.
Color can give us information about the structure of atoms. Now we shall go on to examine the relationship between color and atomic structure in more detail.

8.3 Energy Levels and Color

Electrons in atoms are not randomly arranged, nor are they arranged in neat orbits the way planets are about the sun. However, they do have a type of order imposed on their motion in the form of energy restrictions. Electrons in an atom may possess only certain energies, and they may not possess any energies between the values dictated by the restrictions. The electrons may make a transition from one of the lower energy states to a higher one, and back again, but transitions to any in-between energy are forbidden. Color arises from these electronic transitions which take place in atoms and molecules.

The fact that electrons are constrained to possess only certain energies is the key to understanding their relationship to color. An electron bound to the nucleus by electrostatic attraction (recall the aluminum disc and the glass rod) can be compared to a baseball. The baseball gets into the air by being whacked with a bat; as long as it is in the air, it carries all the energy of the whack with it. An electron moving around the nucleus is like the ball in the air; it always has a fixed amount of energy. However, whereas the baseball can be have any amount of energy, depending on how hard it is hit, the electron is restricted to only certain energies. Transferring this idea to the baseball analogy, it is as if the baseball can accept only, let's say, three possible whacks from the batter, one that gives it enough energy to get as far as first base, one that can get it to second base, and one that can send it into the bleachers. If the batter tries to whack the ball anywhere else, it goes automatically into the catcher's mitt. Likewise, the electrons in atoms can have only certain energy levels (batting distances) and nothing in-between.

The energy levels of the electron in an atom can also be compared (but very limpingly - all analogies are imperfect) to books in a bookcase. Each book must occupy a certain shelf (energy level). It would be strange indeed to see a book suspended between shelves. It takes more energy to lift a book to a higher shelf, so the higher the shelf (energy level), the more energy the books on that shelf have by virtue of their position. Now, although electrons may occupy only certain energy levels, they may certainly change levels, just as books in a bookcase may be changed from one shelf to the next. If I pick up a book on the second shelf and move it up to the third shelf, I have imparted enough energy to the book to allow it to occupy a higher energy level. If I allow the book to fall back to the second shelf, it must give up its newly acquired energy in the form of sound and mechanical energy. Similarly, electrons can move from one energy level to another. Consider the hydrogen atom, for example. I expect that the single electron in the hydrogen atom will occupy the lowest possible energy state. This state of the atom is called the "ground state" (remember the baseball analogy). If I give the hydrogen atom an "atomic hotfoot," that is, if I impart energy to it in the form of heat or electromagnetic radiation, the electron in the ground state can make a transition to a higher energy state by absorbing some of that energy, and the whole atom is now said to be in an "excited state." This move, of course, will leave a vacancy in the ground state level and so you might expect that the electron will return to the ground state at the earliest opportunity. In so doing, the electron has to give up its newly acquired energy, and it does so by spinning it off in the form of electromagnetic radiation.

The color of the emitted electromagnetic radiation, *i.e.,* the frequency, depends upon the difference in energy between the two energy levels. Recall, from the discussion of wave motion in Chapter 3, that the frequency of a wave is directly proportional to its energy. The frequency of the light emitted when an electron makes the transition back from the higher to the lower energy level depends entirely upon the difference between the two levels.

8.4 The Energy Levels of the Hydrogen Atom

The electronic energy levels of the hydrogen atom have been the most intensively studied because hydrogen, with only one electron, is the simplest case of this type. Figure 8.1 is an energy level diagram which shows some of the possible energy levels for the hydrogen atom and some of the transitions to higher energy states and return to lower energy states. These values were obtained by spectroscopists from the total spectrum of energized hydrogen atoms including ultraviolet, infrared and X-ray regions of the spectrum. Each transition from a higher energy level back to a lower one involves the release of energy in the form of radiation, and since the energy is discrete, the radiation associated with it will be observed in the hydrogen spectrum as a narrow line, just as you observed line spectra in Miniexperiment VIII. However, since some of these transitions are undetectable by the human eye, they must be recorded by special instruments. The transitions in the infrared region can be recorded by an infrared spectrophoto-meter, and the transitions in the ultraviolet can be recorded by an ultraviolet spectrophotometer. A spectrophotometer is simply a device which can measure (meter) the radiation (photo) emitted over a certain region of the spectrum (spectro). The colored lines, of course, will be visible to the eye.

In Figure 8.1, the vertical axis represents units of energy in electron volts (eV). You will notice that the energy levels are not evenly spaced, which means that the energy level differences between states are not the same. Table 8.1 lists these energy differences for hydrogen, as well as the energy differences associated with some of the transitions from level to level. The energy levels are given numbers from 1 to 6 in order of increasing potential energy, relative to the nucleus. The list, however, could have been extended from 1 to infinity.*

The transitions listed in Table 8.1 fall into certain groups or series named after the spectroscopists who first described each of them. The second, or Balmer, series involves transitions from the second energy level (the first-excited state) to higher energy levels and back again. Notice that when excited electrons return to the second energy level, they lose energy between 410.29 and 656.47 nm. These energy transitions correspond to wavelengths in the visible region of the spectrum (See Table 3.1), so we expect to be able to see the Balmer series transitions with our own eyes. Furthermore, these transitions correspond to the emission of violet, blue and red light, so if we use a dispersing apparatus like a prism or grating, we should expect to see these colors when the hydrogen atom electrons fall into the first-excited state. The other transition series may also be taking place simultaneously in other atoms but since they fall into the ultraviolet and infrared regions, they cannot be seen by the naked eye. (A word of caution: ultraviolet radiation is highly energetic and, although it is not visible, it is certainly present and can cause damage to the eye. Therefore, these spectra should never be viewed without protective goggles which filter out the ultraviolet radiation.) Taken together, all of these spectral transitions can be recorded on photographic film, and a film record serves to identify the presence of hydrogen. This fact alone has proved invaluable to astronomers who have identified hydrogen in this way to be the most abundant element in the universe.

---

*An electron removed to an infinite number of energy levels away from the nucleus constitutes effectively complete removal of the electron. The remaining atom now has an excess positive charge and is called a positive ion, or cation.

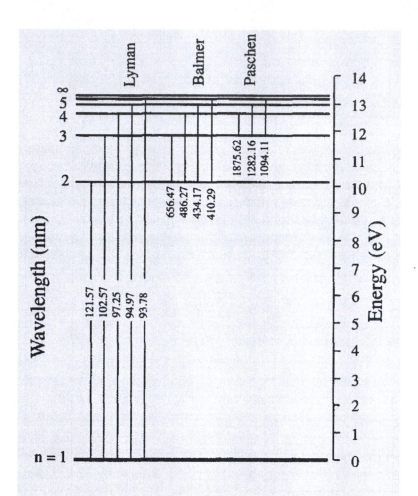

FIGURE 8.1. Electronic Energy Levels of
Hydrogen

ENERGY LEVEL DIFFERENCES IN THE HYDROGEN ATOM

Series	Transition	Energy (eV)	Frequency (sec^{-1} X 10^{15})	Wavelength (nm)
Lyman	1 ↔ 2	10.208	2.468	121.57
	1 ↔ 3	12.098	2.925	102.57
	1 ↔ 4	12.759	3.085	97.25
	1 ↔ 5	13.065	3.159	94.97
	1 ↔ 6	13.231	3.199	93.78
Balmer	2 ↔ 3	1.890	0.457	656.47
	2 ↔ 4	2.552	0.617	486.27
	2 ↔ 5	2.858	0.691	434.17
	2 ↔ 6	3.023	0.731	410.29
Paschen	3 ↔ 4	0.662	0.160	1875.62
	3 ↔ 5	0.968	0.234	1282.16
	3 ↔ 6	1.133	0.274	1094.11
Brackett	4 ↔ 5	0.306	0.074	4052.27
	4 ↔ 6	0.471	0.114	2625.87
Pfund	5 ↔ 6	0.165	0.040	7459.85

To recapitulate: when an electron bound in a hydrogen atom absorbs energy, it is constrained to accept only those energies corresponding to transitions to higher energy states if it is to remain within the atom. Referring to Figure 8.1, an electron in the lowest energy state (the ground state) can only take up energies of 10.208, 12.098, 12.759, *etc.*, electron volts. The electron cannot accept energies between these values; it must occupy only certain energy states corresponding to the levels shown in the diagram. If it occupies state 2, it is one energy level above the ground state and is said to be in the first-excited state. If it occupies state 3, it is in the second-excited state, and so forth. When it de-excites, that is, when it emits this excess energy, it can only move to a lower energy level; it cannot move to a position in-between energy levels and therefore, it can only emit certain energies. For example, an electron in level 3 can only move to either level 2 or level 1 and emit 1.890 eV (656.47 nm) or 12.098 eV (102.57 nm) respectively.

So far, we have limited our discussion to hydrogen, which has a single electron. However, you might expect that the more electrons an atom possesses, the more electronic transitions are possible. Elements like gold, silver and iron can produce line spectra containing thousands of lines. Since each element emits a spectrum which is unique to itself, the spectra can act as a set of "fingerprints" for the purpose of identifying the presence of these elements in various substances. Atomic spectra can be used not only to identify the element qualitatively, but also to determine how much of that element is present in the sample. This technique has been utilized in all fields, from astronomy to police science. It finds great use in art conservation and museum science as well.

In Miniexperiment VIII, instructions were given for observing the line spectra of elements. For example, if lithium chloride were heated in a Bunsen burner flame, energy would be imparted to some lithium atoms to cause electronic transitions as lithium "relaxes" back down to lower energy states and emits energy as a bright, crimson flame. This is because one of the major energy transitions in lithium corresponds to a wavelength of 670.8 nm and this occurs in the red region of the spectrum. Other elements exhibit flames of different colors corresponding to their most probable transition in the visible region. Barium, at 553.6 nm, gives a beautiful yellow-green flame, and strontium, at 681 nm, gives a red flame which is visually distinguishable from that of lithium.

Before going on to discuss how substances modify the light that shines on them, there is one further topic on the internal structure of the atom which deserves attention.

8.5 Electron Motion and Position

So far, we have discussed only the energy levels of atoms and energy transitions. We have said nothing about where an electron is located when it is in an excited state, as compared to the ground state, but we did hint at the fact that "energy level" and "location" are somehow related.

You know that electrons must be attracted to the positive nucleus, so that they should be found in the vicinity of the nucleus. You might even expect that the electrons could be found at the nucleus. Let us look again at the simplest atom, hydrogen, and see what we can learn about this one-proton, one-electron system.

If I were to place a single proton (= hydrogen nucleus) in free space and then allow a single electron to approach it from a very large distance, I would probably expect that once the electron found itself within the force field of the proton, it would be accelerated toward it and would eventually "fall in" to the proton because of the very strong attractive forces between the two particles. This would be analogous to the north and south poles of two magnets being drawn together spontaneously once they were close enough to each other. In actual fact, the electron is strongly attracted to the proton and once it comes close enough to the proton, it becomes bound to it by an attractive force due to the opposite charges each possesses. However, it is very difficult, indeed impossible, to pinpoint the position of the electron within the atom. Scientists realized very early in their investigations into the atom that the only way to handle a problem like this is to shift our attention from electron position to electron probability. The question now becomes: How does the probability of finding the electron in the atom change as we move away from the nucleus? From our previous discussion, for the single electron of hydrogen in its unexcited state, the electron's highest probability is at the nucleus itself. Electron probability drops off precipitately as we move away from the nucleus, but no matter how far away we may move, the probability never drops to zero. A graphical representation of this reality is found in Figure 8.2. The vertical axis is labeled "probability," a statistical term which can be measured in percentage. The horizontal axis is "distance from nucleus," and is measured in a convenient distance term such as nanometers or Ångstroms. The figure shows that electron probability increases as one approaches the nucleus and that as the distance from the nucleus increases, electron probability decreases rapidly.

This probability picture makes it very difficult to define the boundaries of atoms. If I define atomic size as the diameter of a sphere which encloses both a nucleus and the electrons associated with it, then the size of every atom, theoretically, is as large as the universe itself since electron

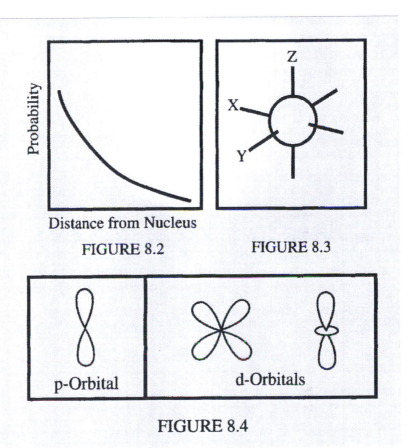

Distance from Nucleus

FIGURE 8.2

FIGURE 8.3

p-Orbital

d-Orbitals

FIGURE 8.4

probability never falls to zero no matter how far away from the nucleus I move. The best that I can do to resolve this situation is to choose a boundary that will enclose 90 to 95 percent of the electron's probability. Figure 8.3 shows a spherical boundary which is meant to depict a 90-95 percent probability of enclosing the electron within the sphere, with the nucleus at the center of the sphere and at the origin of the coordinate system. Some electrons have boundary contours which are not spherical, and some of these shapes are depicted in Figure 8.4. Electrons with spherical symmetry are called "s" electrons. More properly, electrons which enter a region in space associated with an allowed energy with respect to the nucleus and which has spherical symmetry are called s-electrons. Other types of electrons are of the "p" type, the "d" type, the "f" type, the "g" type and so forth.

Although the concepts already introduced are sufficient for what we must do next, it should be mentioned at this point that a more convenient description of electron probability takes into account the increasing volume an electron can occupy in each tiny increment of spherical volume with increasing distance from the nucleus. To understand this last statement a little better, let us imagine that the atom is an onion. At the center of the onion is the nucleus. Although the probability of an electron being found at the nucleus is great (for s-electrons), the volume of the nucleus is very small. Multiplication of this probability (large number) by the volume factor (very small number) yields a very small number. Moving to the first onion layer beyond the center, electron probability drops off for a given point within that volume, but there are many more points to look at because the volume involved is so much larger. Multiplication of the smaller electron probability by the larger volume factor yields a larger number than in the first instance. If we continue to carry out this multiplication for each succeeding layer of the onion, we find that the volume factor continues to contribute to an increase in finding the electron within it, but that the decreasing electron probability makes the product of the two smaller. At some point along the way, the product of the two reaches a maximum, and then falls off to very small numbers as distance from the nucleus increases. For the hydrogen atom, the maximum occurs at 0.059 nm from the nucleus, and a sphere drawn at this position from the nucleus defines the position where the electron is most likely to be if I take into account the entire volume of the atom. Figure 8.5 shows the occurrence of the maximum, and Figure 8.6 depicts the electron pinpointed at one location on the surface of the sphere. Although Figure 8.6 might make you think that an atom resembles the solar system with the nucleus the sun and the electron a planet, be careful. This is not the case. The electron may be found anywhere in space. It is most likely to be found at this given distance on the surface of an imaginary sphere whose radius is the distance.

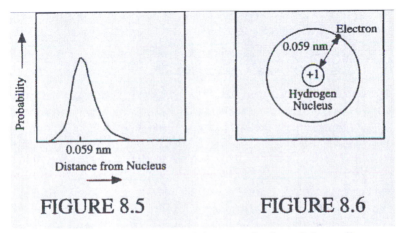

FIGURE 8.5   **FIGURE 8.6**

To give you an idea of the magnitudes of these distances within the atom, we must calculate the nuclear radius. Nuclear physicists have succeeded in approximating this value for hydrogen and it is taken to be 1.4×10^{-6} nm, or 0.0000014 nm. This is such a small radius that an electron 0.059 nm away from the nucleus is about 42,000 nuclear radii away! To put this in a mind's eye scale for you, imagine that we could scale the hydrogen nucleus up to the size of a small peach or a plum (3 to 3.5 cm in radius). On this scale, its associated electron would most likely be found nearly a mile away. Obviously, Figure 8.6 was not drawn to scale.

Perhaps at this point you are wondering what this lengthy discussion has to do with color. Color arises from atoms, and more specifically, from energy transitions that take place in atoms and molecules. In order to understand the nature of these transitions, we must have some idea of the structure of the atom, but it is important that you not have an oversimplified view of atomic structure because then you will be at a loss later on when we have to look at "delocalized" electrons. It is essential that you realize that the best we can do with electrons is to describe them in terms of finding them in a particular location. But, location or position is not the whole story. You also know that an electron in a particular energy level has a certain amount of energy associated with it, and that no matter how far or how near the electron is to the nucleus, as long as it remains within its given energy level, it has a constant energy with respect to the nucleus. However, it also occupies a region in space with certain boundary contours. Such a region is called an "orbital," and the electrons which occupy the s, p and d-shaped regions in space shown in Figures 8.3 and 8.4 are said to occupy s-, p- and d-orbitals respectively. We shall see in future chapters that the presence of such electrons will be very much responsible for observed color in certain chemical compounds.

8.6. Selected Readings

Jordan, Thomas M. Spectacular Spectroscopy. *The Science Teacher* **1989**, *56* (No. 4), 38-40.

Sacks, Oliver A Pocket Spectroscope. *Chemical Heritage* **2001-2002**, *19*(4), Winter, 9, 40-43.

Chapter 9

Colored Objects

The purest and most thoughtful
minds are those that love color the most.

John Ruskin,
The Stones of Venice, II, Ch. 5

Full many a gem of purest ray serene
The dark unfathom'd caves of ocean bear....

Gray's "Elegy Written in a Country Churchyard"

CHAPTER 9

COLORED OBJECTS

9.1 Why Is the Book Blue?

In Chapter 6, we spent a good deal of time discussing some sources of light and we made the distinction between various sources on the basis of their spectral energy distribution curves. Each of the sources we examined, the sun, tungsten illumination, fluorescent illumination, *etc.,* was its own source of lighting. Such a source is called a luminous body. Other examples of luminous bodies are stars, red-hot coals and welders' arcs. In the present chapter, we shall examine the interaction between luminous bodies and the objects which they illuminate. The page you are reading right now is an example of an illuminated body. It is visible only by virtue of the radiation from a luminous body falling upon it. If the luminous body is removed, the illuminated body is no longer visible. One of the most striking examples of an illuminated body is the moon. The moon is visible only by virtue of its illumination by the sun. It is important to distinguish carefully between luminous and illuminated bodies from now on.

In the preceding chapter, we confined our discussion of hydrogen to the "atomic hotfoot" approach. Let us take a gentler approach and see if it yields any more information. Suppose that instead of bringing out the heavy artillery and hitting hydrogen with an overwhelming amount of energy such as a very hot flame, we allow a beam of electromagnetic radiation ranging from 50 to 700 nm to traverse the sample of hydrogen. Since this radiation contains all wavelengths of light between the given limits; it certainly contains the wavelengths corresponding to the transition energies of the Lyman and the Balmer series, namely 93.78 nm, 94.97 nm, 97.25 nm, 102.57 nm, 121.57 nm, 410.29 nm, 434.17 nm, 486.27 nm and 656.47 nm. As these wavelengths enter our collection of hydrogen atoms, they and only they will be capable of bringing about the electronic transitions proper to hydrogen. All other wavelengths are of either too great or too little energy to accomplish this. These wavelengths are called the resonance frequencies or resonance wavelengths of hydrogen for these series. Only these wavelengths are capable of being absorbed by the hydrogen atoms. Therefore, as these waves traverse our sample of hydrogen, the hydrogen atoms will select out these wavelengths and allow the others to pass on through undisturbed. As the light beam emerges from the other side of the hydrogen sample, it will be missing the wavelengths of the Lyman and Balmer series. If I were to take the emergent beam and pass it through a prism or grating in order to disperse the light of varying wavelengths, there would be gaps, or dark lines, where the missing wavelengths should have been. The positions of these dark lines in the spectrum are well-known for hydrogen, and it is this specific method that astronomers use to identify hydrogen in outer space. Furthermore, every element has its own specific dark-line spectrum which can act as a ready-made fingerprint for that element. (Perhaps you have figured out by now that the dark-line spectrum of an element is the "negative" of its bright line spectrum.)

Within each element's own particular set of energy transitions, some transitions are more likely to take place than others. In other words, each transition itself has a certain probability associated with it, and these probabilities can be calculated by applying equations which arise from modern quantum theory. Without trying to undertake this complicated task, it is possible to observe the most probable transition for certain elements in the visible region with our naked eye. This is the basis of what are known as "flame tests" for the various elements, and the reason for the colors of the lithium, barium, sodium and other metallic salts mentioned in Chapter 8 when they are placed in a hot flame.

Now, let us take a more complicated system than hydrogen and see if we can apply some of the principles we learned in dealing with the simpler system. If I take a glass of water and examine it in sunlight, I notice that most of the light can traverse the water and emerge from the other side of the glass. This behavior is known as transmission. However, I also notice that some of the light is reflected back to my eyes because I may be able to see my own face in the glass. This behavior is called reflection. However, it is not immediately apparent if all of the light striking the glass is either transmitted or reflected. Some of it may be absorbed, just as our hydrogen sample absorbed light and ultraviolet radiation.

To observe absorption in our glass of water, let us add a drop of blue vegetable dye to the water and mix it up. Now as I look through the glass, I notice that the only color emerging from the other side is blue, whereas previously, all of the colors, that is, white light, were transmitted. Now only blue light is being transmitted. If only blue light is being transmitted, then the rest of the light must have been soaked up or absorbed by the blue dye. This corresponds to our hydrogen sample absorbing certain wavelengths of radiant energy to produce a dark line spectrum. However, the blue vegetable dye is a much more complex system than hydrogen atoms; in fact, the dye can undergo so many more electronic transitions than hydrogen that the dye absorbs large portions of the visible spectrum. In fact, if it leaves blue light unabsorbed, then it must have absorbed a good deal of the red, orange, yellow and

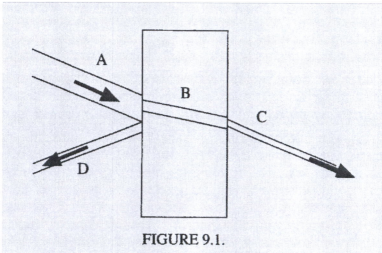

FIGURE 9.1.

A. When light falls upon an object,
B. part of it may be absorbed and refracted,
C. part of it may be transmitted if the object is transparent,
D. and part of it may be reflected.

green regions of the spectrum. This phenomenon is called selective absorption; it is responsible for the color we see in transparent and opaque objects.

So far, we have observed the glass containing the blue dye with sunlight, which contains all of the colors. Let us now take a prism, disperse the sunlight into its constituent colors, and then allow each color in turn to fall upon the glass. If I allow the blue-violet end of the spectrum to impinge upon the glass, I notice that most of the light is transmitted; I see a ray of blue-violet light emerge from the other side of the glass. However, as I try each of the other colors in turn, I notice that very little green light is transmitted, still less yellow and orange and practically no red light. These colors are all absorbed to a large extent by my blue object, and as a matter of fact, this is the reason why the object is blue.

Next, let me examine a blue book in the same way. This time, I notice that none of the light striking the book is transmitted; the book casts a sharp shadow, and I term it "opaque." However, some of the light must be reflected since the book is visible, and since it appears blue, it must be the blue portion of the spectrum that is being reflected to my eye.

We have mentioned several ways in which light can interact with matter: reflection, transmittance and absorption, and these are summarized by Figure 9.1. Let us now look at each of these modes of interaction in more detail.

9.2 Reflection

We all know that a light beam can be modified with respect to the direction in which it travels by being reflected from the surface of an object. In certain special cases, not only the appearance of an object, but also its color, can be affected by the manner in which it reflects light. In most instances, our experience of reflection from a surface involves our observation of reflection from a smooth surface such as a mirror or a pool or a polished metal surface. In these cases, we always expect to see some sort of reflected image, and the integrity of the image depends upon the smoothness of the reflecting surface. The reason why this is so is that light emitted by an object or a source as parallel rays will be reflected from an object at an angle equal to the angle of incidence. Reflection from a smooth surface is illustrated in Figure 9.2. In this case, the reflected rays all remain nearly parallel because each ray has a normal plane which is parallel to the normal planes of all the other rays.

In the case of reflection from a rough surface, on the other hand, normals must be constructed perpendicular to the surface which a particular ray is striking, and these are seldom parallel, as illustrated in Figure 9.3. Since the reflected rays are no longer parallel either, very few rays from the same surface region will reach the observer's eye, so the surface will appear quite dull. This phenomenon explains the difference between a glossy finish and a matte finish. Furthermore, if the surface is very rough, it is possible that hardly any reflected light at all will reach the observer.

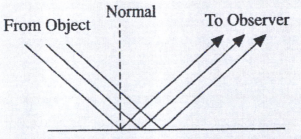

FIGURE 9.2. Reflection from a Smooth Surface

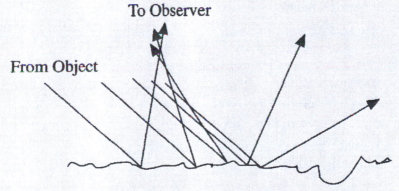

FIGURE 9.3. Reflection from a Rough Surface

9.3 Transmittance

As the word implies, transmittance occurs when light strikes an object and some of the incident light emerges from the other side. If the light beam strikes the object "head-on," that is, at an angle of $90°$ with the surface of the object, the light beam will coincide with the normal plane and the angle of incidence will be $0°$. In this one case, the angle of incidence will equal the angle of refraction and, although the light beam will change its velocity, its direction will remain unchanged. A small amount of the light will also be reflected from the object, and if an observer happens to be standing between the light source and the object, the light will be reflected to the observer, rendering the object "visible." This is illustrated in Figure 9.4. Of course, if the light beams striking an object are not at an angle of $90°$ with the surface, then refraction will take place as well, as illustrated in Figure 9.5.

When light of any wavelength range is transmitted by a medium to a great enough degree that we can clearly see an object through it, the medium is said to be transparent. Light rays come from the object in an orderly array, pass through the transparent medium with this arrangement unchanged (or shifted uniformly by refraction), and emerge to reach the observer in the same orderly array. If the surfaces of the medium are uneven or scratched, or if the medium has small foreign particles in it that scatter the light, then the medium may still transmit the light, but the orderly array is destroyed, and no image can be discerned through the medium. The medium is then said to be translucent. An example of a transparent object is a pane of window glass; an example of a translucent object is a glass brick with a scored or uneven surface.

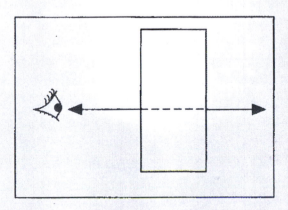

FIGURE 9.4. Reflection at an Angle of 90º
to the Surface

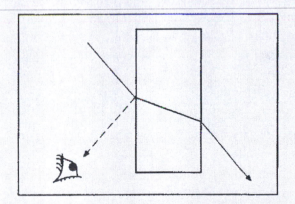

FIGURE 9.5. Reflection and Refraction at
an Acute Angle to the Reflecting Surface

9.4 Absorption

In many cases, incident radiation containing all of the wavelengths of visible light is absorbed by matter in relatively large amounts and over a relatively large wavelength range. We saw in Section 9.1 that a dark line spectrum could be produced by an element which selectively absorbed certain wavelengths of incident radiation. A more complex absorbing material could conceivably absorb not merely a selected wavelength here and there, but whole ranges of wavelengths. The spectrum of such an absorbing material would show dark bands rather than dark lines. Figures 9.6 a and b illustrate the continuous visible spectrum, representing the waves of visible light emitted from an incasdescent source such as a tungsten lamp; none of the frequencies of light have been absorbed. Figure 9.6 c represents the spectrum that would arise if the incandescent lamp spectrum were passed through hydrogen gas. The dark lines, from left to right, are the wavelengths absorbed by the gas and correspond to the Balmer series lines 410.29, 434.17, 486.27 and 656.47 nm respectively (See Table 8.1). Figure 9.6 d represents the spectrum that would result if a sample of pure hydrogen were subjected to a high voltage discharge: the electrons promoted to higher energ states undergo transitions back to lower energy states and emit in the visible region the corresponding bright lines of the Balmer series in the process. Figure 9.6 e represents what happens to the spectrum of an incandescent lamp when it impinges on, not a simple atomic gas, but a very complex molecular system which can undergo many electronic transitions in the visible region. In this case, the complex system undergoes numerous transitions in the red-orange-yellow and the blue-violet regions, but hardly any in the blue-green-yellow region of the spectrum, which remains unabsorbed. For this reason, the object appears green. As a matter of fact, this is the "band spectrum" of grass.

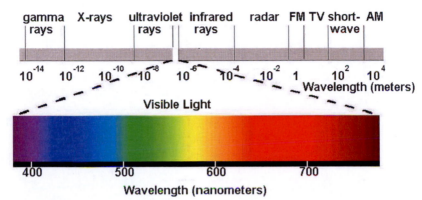

Fig. 9.6a The Electromagnetic Spectrum (visible spectrum highlighted)

Fig. 9.6b Continuous Spectrum of an Incandescent Source

Fig. 9.6c Dark-Line Spectrum of Hydrogen

Fig. 9.6d Bright-Line Spectrum of Hydrogen

Fig. 9.6e Band Spectrum of a Green Object (Grass)

9.5 Why Objects Appear Colored

All non-luminous objects which are colored appear so because they absorb part of the incident radiation and reflect or transmit the remainder to the observer. The colors seen by the observer are the ones that remain unabsorbed.

(a) Transparent Colorless Objects. A transparent object is colorless when it is capable of transmitting all wavelengths of the visible spectrum. The visible light which shines on such an object emerges unchanged in color. Ordinary window glass, Lucite™ and water are examples of transparent, colorless materials. Figure 9.7 illustrates the behavior of a prismatic spectrum when it encounters a colorless, transparent object.

(b) Transparent Colored Objects. Colored transparent objects transmit only certain wavelengths of visible light and absorb the rest. The color seen emerging from the other side of the object is the color of the object itself. The color of the object is the color of the light it transmits because the other colors are absorbed by the object. For example, a piece of red glass illuminated by white light will transmit red light (and perhaps some orange and yellow), but absorb all others, as illustrated in Figure 9.8. Furthermore, the appearance of the object depends very much on the type of light falling on it. Suppose, for example, I were to take the beam of red light transmitted by the object in Figure 9.8 and shine it on a blue transparent object such as a piece of blue cellophane. Since the cellophane is blue because it transmits only blue light, then it will surely not transmit red light. Therefore, in pure red light, the object will transmit no light and appear black and opaque.

(c) Opaque White and Black Objects. An object appears white when it mainly reflects all of the wavelengths of visible light. The reflection may be accompanied by some scattering and a

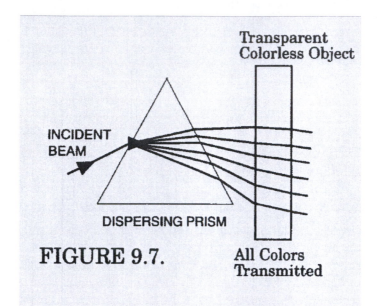

FIGURE 9.7.

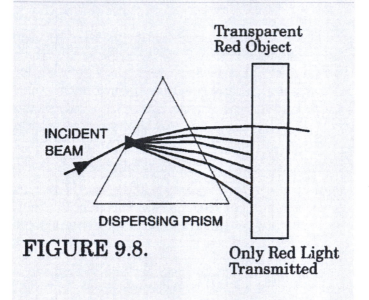

FIGURE 9.8.

small amount of absorption, but no transmission. The object will be seen by virtue of the reflected and scattered light, and will appear white and opaque.

It is important at this point to distinguish between white and colorless objects. Colorless objects are colorless because they transmit almost all of the wavelengths of visible light. For this reason, colorless objects are also transparent. A common example of a colorless object is ordinary window glass. A white object, on the other hand, reflects and scatters most of the visible light striking it. For this reason, white objects are also opaque; they cannot be seen through. A common example of a white object is a sheet of ordinary typing paper.

An ideal opaque white object may be defined as one which reflects all wavelengths of visible light 100 percent. Of course, no such object exists since some absorption and scattering of light always takes place, but some metal oxides such as magnesium oxide have been prepared with surfaces which produce a very high degree of reflectance and are used as reflectance standards in color measurement.

Since black can be defined as the opposite of white, then an ideal black object is one which absorbs all visible light. Again, since most black objects reflect and scatter a little of the light, they are not ideal, and are therefore visible by virtue of the light they scatter as well as by virtue of the un-blackness of the background. All black objects are opaque to visible light.

When light is absorbed by a black object, it is converted to other forms of energy, including heat energy. Benjamin Franklin performed a famous experiment by laying out large pieces of black cloth and white cloth over snow on a sunny day. While the black cloths sank deeply into the snow as the snow beneath them melted, the effect was much smaller for the white cloths which largely reflected the sunlight instead of converting it to heat. Based on this experiment, Franklin recommended that people wear white clothes in the summer and dark clothes in the winter. Ever since then, fashion has followed Franklin's suggestion by decreeing summer clothes of pastels and whites and winter clothes of brown, black, navy, rust and dark green.

Is gray a color? You know that gray is produced by mixing black and white. Gray, therefore, is the result of partial absorbance and partial reflection of all the visible wavelengths. Various shades of gray are possible by varying the proportion of black to white.

Keep in mind that the quality we call opacity changes as the wavelength of illuminating radiation changes. The transparent red object of Figure 9.8 is transparent only to red light; it is opaque to all other wavelengths in the visible region. However, any object may be transparent to other regions of the electromagnetic spectrum but opaque to visible light. For example, brick walls and human flesh are opaque to visible light, but are transparent to X-rays and gamma rays.

(d) Opaque Colored Objects. You recall that if a transparent object is red, it is red because all other visible wavelengths are absorbed. The red light, which traverses the object largely unabsorbed, reaches your eye. You have probably surmised by now that opaque red objects are red because they also absorb almost all of the visible wavelengths except red, and that the red light is reflected back to your eye. In other words, an object, whether opaque of transparent, has the color it has because the color you see is what is left after all the other colors have been selectively absorbed. The color you see is the "rejected" color, if you will, and you would not see it if it had not been rejected because it would have been gathered into the object itself and would not have been able to reach your eye. The color that reaches your eye only does so because it is not hindered from doing so; the colors that you do not see are those that are absorbed by the object.

Miniexperiment IX

The human brain interprets an object as gray by comparing it to other objects. Thus it is possible to make a gray color look white or black to the eye.

Take a piece of black matte paper to a darkroom and pin it up to the wall. Turn off the light, wait until your eyes are accustomed to the dark, and then shine a flashlight onto the black paper. The black paper is not a perfect 100% absorber of light, so it will actually appear gray since it is the brightest reflector visible. Can you devise a method for making a gray object look white?

Draw different colored squares on white paper using colored chalk or paint. Now, hold pieces of cellophane or light filters of different colors over each square. How does the filter affect the color? Now try drawing a picture using the colored chalk or paints so that the observer sees one thing in white light, but sees a different image when one of the colored filters is held over it.

It is interesting to note that the color-producing mechanism for metals is opposite to that for nonmetals. The commonly observed colors of metals are so-called "surface colors" insofar as they result from the selective reflection of various wavelengths and that this reflection is highest at wavelengths where absorption is also highest. This phenomenon is possible for metals because of the fact that electrons in metals have higher mobility than those in nonmetals. (Remember that metals are good conductors of electricity.) When a light wave strikes an object, it will interact with that object's electrons, and the greater the mobility of the electrons, the greater the degree of interaction. At the same time, the higher mobility of free electrons results in much higher reflectance values for metals. For example, nonmetallic, crystalline titanium dioxide has a surface reflection of about 21% in air, but polished aluminum reflects over 90%. Metallic-like reflection can occur with some nonmetals possessing a high degree of electron mobility. This gives rise to the "bronzing" effect known in the coatings and ink industries.

9.6 Lambert's Law and Beer's Law

Two laws of importance to visual artists both apply to transparent materials. Since no transparent materials are ideal, they all absorb some of the incident radiation to some degree. Consequently, if the material were to be made sufficiently thick, eventually no light at all would pass through. A room walled by fifty-foot thick glass walls can be completely dark on a sunny day. Similarly, virtually no light reaches the bottom of the deepest ocean trench even though a glass of seawater is quite transparent.

The thicker the transparent medium through which light passes, the less light is transmitted. If a block of glass one inch thick transmits 80% of the incident light, a second one inch thick block placed behind it will transmit 80% of the 80%, or 64% of the incident light. Further calculation shows that it takes 4 thicknesses of the glass to reduce the intensity to less than 50% of the original intensity, and 11 thicknesses to reduce the intensity to less than 10% of the original. However, theoretically, the intensity of the transmitted light should never fall to zero. The mathematical expression of this observation is known as Lambert's Law. The law must be applied separately to each wavelength of light since some transparent objects transmit greater intensities of some wavelengths than others.

A variation of this same effect is called Beer's Law and deals with colorants in solution. It is the colorant such as a food dye dissolved in water which absorbs one or more of the colors of incident light and transmits the remainder. Thus, the more colorant in a given quantity of solvent, *i.e.,* the more concentrated the solution, the more light is absorbed. This is known as Beer's Law when stated quantitatively: equal relative amounts of absorption occur when light passes through equal amounts of colorant. As with Lambert's Law, this is true for only one particular wavelength, so measurements must be made for transmittance at each wavelength.

Beer's Law may be observed in two ways, either by passing light through different thicknesses of colored solutions, or by passing light through different concentrations of a colored solution with a fixed thickness. Since Beer's Law and Lambert's Law are so closely related, the mathematical expressions that govern each are usually combined to yield the Beer-Lambert Law.

Let us see how Beer's Law works. A solution which appears red absorbs most of the incident light except for the red wavelengths. The larger the container, the darker the solution will look since more of the incident light is being absorbed. If a red dye solution transmits 50% of the incident light per inch of solution, then two inches will transmit 50% of the 50%, or 25%, and 4 inches will transmit 50% of the 50% of the 50% of the 50%, or only 6.25% of the incident light. The solution would probably appear almost opaque and a deep orange-black. In the same manner, increasing the concentration of a dye will give the same results. Both situations are illustrated in Figures 9.9 and 9.10.

Sometimes Beer's Law can operate to cause a color change. This is an effect that may occur not only with solutions,

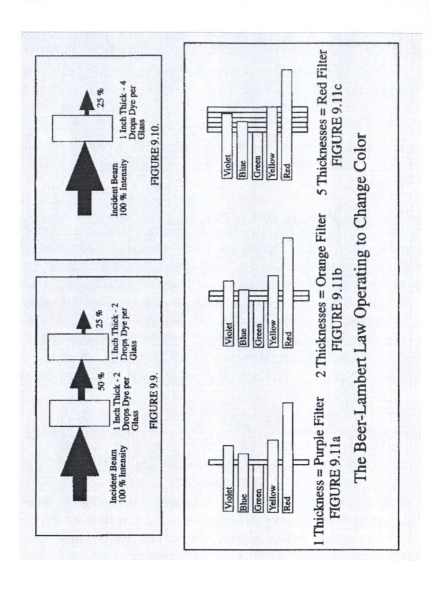

Incident Beam 100 % Intensity — 25 %

1 Inch Thick - 4 Drops Dye per Glass

FIGURE 9.10.

Incident Beam 100 % Intensity — 50 % — 25 %

1 Inch Thick - 2 Drops Dye per Glass

1 Inch Thick - 2 Drops Dye per Glass

FIGURE 9.9.

Violet, Blue, Green, Yellow, Red

1 Thickness = Purple Filter
FIGURE 9.11a

2 Thicknesses = Orange Filter
FIGURE 9.11b

5 Thicknesses = Red Filter
FIGURE 9.11c

The Beer-Lambert Law Operating to Change Color

but also with certain water colors. Water colors are transparent paints which are reflected to the viewer's eye from a white ground. It is evident from the preceding discussion that the thicker the paint layer, the more opaque and more dark a water color becomes, but it may not be so obvious that the color can also change. Let us see how this can happen.

Examine a purple filter. The purple is the color recorded in our brain for the combination of red, yellow and some blue-violet wavelengths which are transmitted preferentially by the filter. For a typical purple filter, about 100% of the red light, 80% of the yellow light, 20% of the blue light and about 40% of the violet light are transmitted. See Figure 9.11 a. If the thickness of the filter is doubled, then by the Beer-Lambert Law, 100% of the red, 64% of the yellow, 16% of the violet and 4% of the blue are transmitted. The yellow-red emergent light is perceived as orange. When five thicknesses of filter are superimposed, the only color that comes through visibly is red, so the color of the five-fold thickened filter is now a faint red. Please see Figure 9.11 b and c.

Although such changes are sometimes pronounced for colored gelatin, plastic or glass, a similar effect can occur with different thicknesses of transparent paints such as water colors. Although manufacturers try to select pigments which are close to pure colors, no "pure color" pigments actually exist. Therefore, artists should test paints for subtle differences in hue as the paint layers are increased.

In the next chapter, we shall examine exactly what is meant by a "spectrally pure" color and see how a pure color can be differentiated from an impure color by measurement of relative amounts of light reflected or absorbed by the colors.

Miniexperiment X

Add a few drops of vegetable dye to about 200 mL of water in a beaker and mix thoroughly. On an overhead projector, if available, place five small (10 mL) beakers Add 2 mL of the dye solution to the first beaker, 4 mL to the second, 6 mL to the third and so on, until the fifth beaker is full of the dye solution. Project the images of the beakers on a screen and estimate the changes in color density in each.

Test the sensitivity of your color vision. Label each of five 250 mL beakers with the numbers 2, 4, 6, 8 and 10 (preferably on the bottom of the beaker). Add water to a depth of one centimeter to each beaker. Add two drops of vegetable dye to the beaker labeled 2, four drops to the beaker labeled 4, and so on. Now shuffle the beakers around so that they are in random order, and try to re-order them. See if your estimated order corresponds to the original order of the beakers.

Take a test tube that is 15 cm (6 inches) long and about 2 cm (3/4 inch) in diameter. Fill it with water and add a few drops of blue food color and mix well. (Baker's food colors work well with this experiment.) Shine a narrow pencil of light from a flashlight covered with a cardboard cone with a small hole cut in the tip of the cone. Start by shining the beam across the 2 cm width of the tube. The beam will appear blue and will be seen as blue in the solution due to dust particles suspended in the water. Now, shine the beam up through the length of the test tube. You will probably be able to observe that the beam changes color as the light proceeds upward so that it is blue at the bottom but red at the top. This effect is similar to that described in the stacking of purple filters. Experiment with different lengths of tube and different quantities and hues of food colors.

Place a colored filter over a flashlight or in front of a lamp covered with a white shade so that you are looking though the colored filter at the light. Now, bend the filter so that a double thickness covers half of it. Compare the double and single thicknesses.

CHAPTER 10
SPECTRAL CURVES

10.1 The Spectrophotometer

As we looked at the causes of color in greater detail in the preceding chapter, we began to speak in terms of relative amounts of light, in percent, being transmitted by transparent objects. In fact, in discussing the changing appearance of a purple filter with increasing thicknesses of material, it became necessary to work with specific percentages of each wavelength range transmitted in order to be able to explain the phenomenon. Although the human eye is a very good judge of color and can judge very nicely between relative concentrations, the eye is incapable of making such judgments quantitative and of making them at each wavelength of visible light. To do this job, an instrument that can disperse visible light into its constituent wavelengths and then measure the amount of light transmitted by transparent objects (or reflected by opaque objects) is needed. Such an instrument is the spectrophotometer.

A spectrophotometer is basically a very simple instrument. It consists, primarily, of what is known as a monochromator. "Mono" means "one," and "chromo" means "color," so basically, a monochromator system is one that (a) can disperse incident radiation into its constituent wavelengths and (b) pass each wavelength, or a very narrow band of wavelengths, into a detection system to be measured. The simplest type of monochromator is a prism and a narrow slit in a piece of cardboard. The prism can disperse visible light, and by slow rotation of the prism, each narrow band of color can be made to fall on a screen behind the slit; if a detection device replaced the screen, we would have a full-fledged spectrophotometer. Modern spectrophotometers are much more sophisticated than the simple one just described, but the basic components are the same. Most spectrophotometers now employ gratings instead of prisms because gratings can disperse radiation more conveniently.

A schematic diagram of a typical modern spectrophotometer is shown in Figure 10.1.

The modern instrument has an elaborate set of slits that allows the experimentalist to vary the wavelength range being recorded, and it also contains special light sources, a drive mechanism to rotate the grating or prism, a sample holder in the path of the beam of light, and a detection system which can convert the light striking it into an electrical signal which is proportional to the intensity of the light. Many instruments also have a dual pathway for the light beam to traverse which allows for the simultaneous comparison of a sample against a standard at every wavelength; such an instrument is called a "double beam" spectrophotometer. Since it is quite tedious to record percentages of transmittance at each wavelength, many instruments also come equipped with a recorder to perform this task. The record, or plot, of Percent Transmittance (or Reflectance) against wavelength is called a spectral curve.

10.2 Spectral Transmittance Curves

The light transmitted by transparent objects can be measured at each different wavelength by means of a spectrophotometer. If we plot the relative amount of light transmitted by the object against the wavelength, we get a spectral transmittance curve. The vertical axis of the curve represents the percentage of the incident light which is transmitted at the selected wavelength. For example,

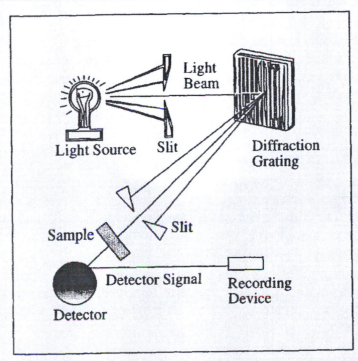

FIGURE 10.1. Simplified Diagram of a Spectrophotometer

if all of the incident light is transmitted by the object at 680 nm, then a point is plotted at the intersection of 100% on the vertical axis and 680 nm on the horizontal axis. Similarly, if only half of the incident light is transmitted by the object at 570 nm, then a point is plotted at the intersection of 50% on the vertical axis and 570 nm on the horizontal axis. Each of the other wavelengths of visible light has its own corresponding percent transmittance for the object in question, and a point by point plot of these values gives rise to a curve similar to that illustrated by Figure 6.1; connection of each of these points yields a continuous curve. Most modern spectrophotometers come equipped with a recorder which produces the continuous curve in a matter of minutes. Built-in microcomputers are becoming standard equipment in many spectrophotometers so that the spectral curve can be recorded and stored digitally as well as produced as hard copy output. Not only that, it is possible to store a reference spectrum that can be subtracted from the sample spectrum, thus eliminating the need for a "double beam" instrument. It is even possible to store multiple scans of the same sample and add them together to produce a composite spectrum for samples that contain very little of the material being analyzed.

Spectral transmittance curves for three gel filters are shown in Figure 10.2 a, b and c. These filters are permeated by dyes which give the desired colors, but there is no way to make a dye which transmits only the wavelengths corresponding to the color green or only the wavelengths corresponding to the color red. Figure 10.2 b is the transmittance curve for a green filter. The range of maximum transmittance extends from about 450 to 630 nm; the remainder of the light is absorbed by the filter. From Table 3.1, we see that the wavelength range corresponding to green light is from 491 to 575 nm, the range within which the transmittance maximum of Figure 10.2 b falls; hence, the color of the filter

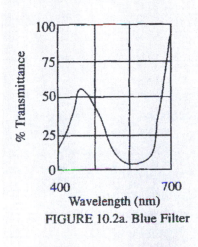

FIGURE 10.2a. Blue Filter

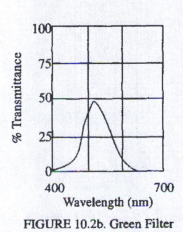

FIGURE 10.2b. Green Filter

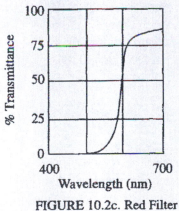

FIGURE 10.2c. Red Filter

appears green. However, there is a slight overlap of transmittance into the blue region to the left of the maximum, and into the yellow and orange regions to the right of the maximum. The transmittance maximum itself occurs at about 530 nm and amounts to 45%.

The effect of viewing the world through a green filter can be understood more readily by recourse to this transmittance curve. When white light shines through a green filter, the curve tells us that only the green portion of the white light manages to get through the filter; the other colors are absorbed by the filter. When the resulting green light shines upon a piece of white paper, the paper appears green since only green light can be reflected back to our eyes from the paper. However, if that same green light were to strike a piece of red paper, the red paper would absorb the green light and would reflect very little, if any, light back to our *eyes*. The color sensation we would experience would be black. What color do you suppose a piece of blue paper would appear when viewed through a green filter?

An understanding of these curves also helps us to describe color in a manner that is independent of the subjective reaction of our eyes. For example, if I were to look at the filter represented by Figure 10.2 a, I would see what I thought was a pure blue color. However, the curve tells me that although the filter allows blue light to pass through it, it also allows some red light to pass through it as well. So little red light gets through that my eye is not sensitive to the "purple" produced but, nevertheless, the blue filter is slightly purple and therefore different from "true blue." And since such descriptions as "true blue" and "off blue" are so qualitative and subjective, it is easy to see that a much more precise system must be adopted in order to describe a color for the purposes of color-matching in commercial as well as artistic enterprises. The key for precise measurement of colors is the spectral curve which gives us quantitative information about the colors we perceive in a qualitative fashion.

Because of the high intensity of the red transmittance of the blue filter of Figure 10.2 a (80% at 700 nm), a red object viewed through this filter will appear red, and a blue one will appear blue, whereas a green one will appear black. Figure 10.2 c is the spectral transmittance curve for a red filter. Referring to Table 3.1 again, what other colors are transmitted by this filter? Would you say that this filter is "pure" red? What color will a blue object appear to be when viewed through this filter?

10.3 Spectral Reflectance Curves

We said in Chapter 9 that a white opaque object reflects most of the incident white light that strikes it. As you might expect, it is possible to measure the degree of reflectance of an opaque object at different wavelengths of light just as we measured the degree of transmittance of transparent objects although, experimentally, it is not always as easy or convenient to do so. The geometry of the instruments used for this purpose must be different from those used for transmittance measurements; a special attachment or a special instrument that can catch reflected light at certain angles must be used. If we were to measure the reflectance of a white object, for example, a layer of finely divided magnesium oxide which has a very high spectral reflectance (about 98%), we would get the curve shown in Figure 10.3. Here we see that all the wavelengths of white light are reflected almost equally and with great efficiency. The ideal white object would reflect 100% at all visible wavelengths. As it is, most white paints give spectral reflectances of about 90%.

What is the color of the object whose spectral reflectance curve is depicted in Figure 10.4?

132

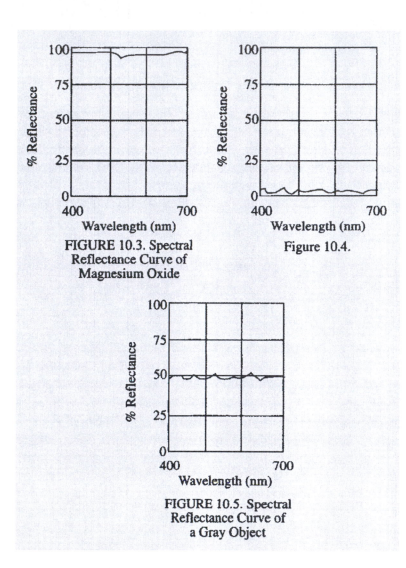

FIGURE 10.3. Spectral
Reflectance Curve of
Magnesium Oxide

Figure 10.4.

FIGURE 10.5. Spectral
Reflectance Curve of
a Gray Object

The spectral reflectance curve of any gray object should show all of the wavelengths of light equally but partially reflected. A medium gray is shown in Figure 10.5.

The spectral reflectance curve for an opaque colored object and the spectral transmittance curve for a transparent object of the same color must necessarily look similar. They can be used interchangeably. For example, a red opaque object colored with the same dye which permeates the gel filter shown in Figure 10.2 c will reflect the same wavelengths of light that are transmitted by the filter. Thus, their curves will look the same, but the vertical axis for the opaque object's curve will be labeled "Percent Reflectance." Such an object will reflect only the wavelengths of light between about 580 and 700 nm; it will absorb all others. Therefore, if I view this object through a blue filter, *i.e.,* if only the wavelengths corresponding to blue light are allowed to strike the object, these wavelengths will be absorbed. The object will reflect very little light back to my eyes, and will therefore appear to be black.

10.4 Additive Lights

Imagine that two projectors are focused on the same spot on a screen. One projector is covered by a blue filter and the other by a green one. These filters have the spectral transmittance curves shown in Figures 10.2 a and b, respectively. Since the amounts of light transmitted by both of these filters are shining on the same spot, the amount of light striking the spot is greater than that coming from either projector alone. Furthermore, the two colors being transmitted add to one another, and the combined color has high intensities of both blue and green in it. Our eyes perceive this color as blue-green; its color name is cyan.

A detailed discussion of this effect is given in Chapter 12.

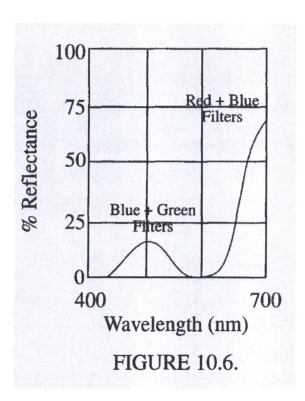

FIGURE 10.6.

10.5 Subtractive Lighting

Miniexperiment XI

Project a spectrum onto a screen by placing a grating in front of a narrow beam of light coming from a slide projector. The narrow beam can be achieved by placing a piece of cardboard with a narrow slit cut in it over the projector lens. Observe the color of a red sheet of paper as it is gradually passed through the spectrum. Do the same for blue and green papers. Now try the same experiment with a yellow paper. Whenever the paper is still colored when illuminated by one of the colors of the spectrum, it reflects and does not absorb that color.

Consider next the case where light is allowed to shine successively though two filters superimposed on one another. We will again use as an example the blue and green filters of Figures 10.2 a and b, but now they will be placed one over the other in front of a beam of light coming from only one projector.

If the blue filter is first in the path of the light, only blue light, plus a little green, violet and red light, passes through the blue filter. This light now strikes the green filter which can, in turn, transmit a high intensity of green light, but only a little blue, yellow and orange light. The result is that both filters together screen out most of the light striking them and allow only a very low intensity of blue-green light to be transmitted. For example, we see from the transmittance curves that at 500 nm, the blue filter transmits 50% of the incident radiation while the green filter transmits 30% of what reaches it; hence only 30% of the 50% or 50% X 0.30 = 15% of the light at 500 nm is transmitted by the superimposed filters. At 700 nm, the blue filter allows only about 10% of the original 62% passing through the blue filter to also pass through the green filter. This amounts to about 6.2% of the incident radiation transmitted at 470 nm. By examining Figure 10.6, you can see that these effects do indeed happen within the limits of error under the conditions of the experiment. Figure 10.6 shows the actual spectral transmittance curve of the white light that has passed through the double layer of blue and green filters. A further discussion of this effect appears in Chapter 11 where subtractive color mixing is discussed in more detail.

10.6 Ideal, Pure and Impure Colors

All the individual colors in the spectrum can be represented by ideal spectral curves. Ideal red will reflect or transmit all of the red wavelengths (about 647-700 nm) 100% and will completely absorb all other wavelengths. Such a curve is shown in Figure 10.7. Ideal yellow will reflect or transmit only the very narrow wavelength band between 575 and 585 nm, and absorb all the rest. The other ideal color curves can be deduced from the information in Table 3.1. They would describe "spectrally pure" colors.

In reality, no pigment or dye gives an ideal curve. If I were to measure the spectral reflectance curve of an ordinary lemon, which I certainly perceive as yellow, I might be startled to find that not only the yellow portion of the spectrum was reflected in high percentage, but also the red, orange and green regions. Thus, a measurement of "lemon yellow" reveals that it is actually a combination of four colors, but is perceived by my senses as the color I call yellow. However, it is not an ideal yellow, nor is it spectrally pure. In this regard, yellow is a unique color. It exists in the spectrum as a spectrally pure color, but strangely enough, our eyes also perceive the combination of red and green light, or red, green and orange light mixed in appropriate amounts as the color yellow. These perceptions can be expressed as follows:

$$Red + Green = Yellow$$
$$Red + Green + Orange = Yellow$$

In fact, a mixture of all the wavelengths of white light minus blue and violet appears yellow: ROYGBV - BV = Yellow

Hence, a filter that transmits yellow light either absorbs everything except yellow, or it absorbs everything except purple and blue, or except purple, blue and orange. Spectral yellow is pure but the yellow produced by all dyes and pigments is impure.

Chemists have identified some pigments which produce perceived colors visually identical to those hypothetically produced by the ideal colors of the spectrum. For example, the dye in the filter whose curve is shown in Figure 10.8 approaches that of spectrally pure red (but not quite). "Tomato red," that is, the spectrum of the compound lycopene, which imparts the red color to tomatoes, contains a good deal of orange and some yellow and is therefore an impure red.

Lest you begin to feel that a knowledge of the spectral curve of a color is the whole story, please do not forget that color depends not only upon the object which modifies the color and gives rise to the spectral curve, but also the spectral power distribution curve of the light source and the nature of the detection system. We have just introduced a detection system in the spectrophotometer that is different from the human eye. Therefore, more than one detection system must be taken into account. Furthermore, when color is examined, it produces its visual effect not only because of its hue, which is measured by the spectral curve, but also by the amount of black and white in the color, and by the amount of the color itself. If color measurement is going to be true to appearance data, then all these factors must be considered.

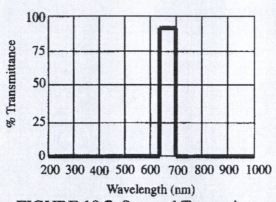

FIGURE 10.7. Spectral Transmittance Curve for an Ideal Red Object

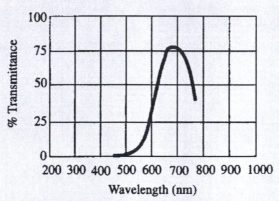

FIGURE 10.8. Spectral Transmittance Curve for a Medium Red Filter (Edmund Scientific Company)

Chapter 11

Subtractive Mixing:
The Colors of Paints

CHAPTER 11

SUBTRACTIVE MIXING: THE COLORS OF PAINTS

11.1 Yellow and Blue Do Not Add to Give Green

When yellow and blue paints are mixed, the product is green. Are you in the habit of thinking that the yellow color adds to the blue color to produce the sum of the two colors? Actually, the color you see is produced in a very different manner; it is the difference between the two colors rather than their sum.

The explanation of the above statement has to do with light absorption. Each pigment is absorbing part of the incident light; we see what is left after each has absorbed its proper wavelengths. The yellow paint is spectrally impure. As we pointed out in the preceding chapter, the color our eye perceives as yellow is the color that absorbs blue and violet light, but reflects the rest of the visible spectrum, *i.e.,* the range between about 490 and 700 nm. Figure 11.1 is the spectral reflectance curve for a yellow paint. The blue pigment is also spectrally impure. It contains a good portion of green if it is the blue commonly described as "artists' blue," and this can be seen in Figure 11.2, the spectral reflectance curve for an artists' blue pigment. If we mix one paint that absorbs blue and violet (yellow paint) and another that absorbs red, orange, yellow and violet, the color remaining unabsorbed in the highest amount by the mixture is green. Hence, this yellow paint and this blue paint mix to yield a green paint. This an example of subtractive color mixing. You can prove this by tracing a copy of Figure 11.2 and superimposing it on Figure 11.1. The area of overlap, mainly in the green region, gives the composite reflectance curve.

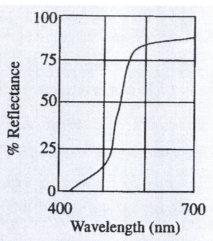

FIGURE 11.1. Spectral Reflectance Curve
for a Yellow Paint

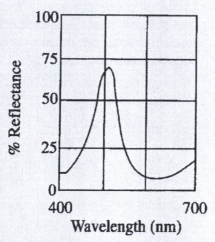

FIGURE 11.2. Spectral Reflectance Curve
for an "Artists' Blue" Paint

11.2 Spectrally Pure Yellow Plus Spectrally Pure Blue Subtract to Give Black

All dyes and pigments in common use are spectrally impure. Combinations of any two spectrally pure colored paints would yield black since each color would absorb all of the reflected wavelengths of the other component. Strange as it may seem, it is only because the chemicals that cause color reflect a variety of wavelengths rather than a single narrow band that color mixing of paints can occur.

11.3 The More Pigments Mixed, the Muddier

A single pigment absorbs a good portion of the visible spectrum, which cuts down on the amount of light available for viewing the pigment. Addition of a second pigment will cause additional absorption of light, and will decrease the amount of reflected light still more. Addition of yet another pigment will cut the reflectance still more, imparting a dull and even "muddy" appearance to the mixture. Thus, it is very important for the artist to achieve the desired color by mixing only two paints. For this reason, it is important to select tube paints that contain only one pigment. Some manufacturers market a range of shades and tints based on different proportions of the same two colors. There is no advantage to their purchase except that of having a desired color pre-mixed. There is a distinct advantage to the artists' purchasing several permanent reds, each of which contains a different pigment. A great deal of experimentation is needed to see how these colors will mix with other pigments. The serious artist will make up a permanent record of a dash of each pure color, and then of various proportions of mixtures. While less spontaneous, it permits greater control. Only when the artist has firm control of his/her tools can freedom and spontaneity be attained.

11.4 Artists' Blue Is Not Blue

Returning to subtractive color mixing, it is a given that commercial blue pigments are not spectrally pure blue, but blue-green in hue. Otherwise, they could not mix with yellow to produce green.

There are, of course, many different greens depending on how much the yellow and blue-green reflectance curves overlap one another. A less precise way of making this statement is that the different shades of green depend upon how "true" the yellow and blue-green colors are. If, for example, the blue-green had more blue than green in it, the subtractive color would be bluish-green when the blue-green was mixed with yellow. Unfortunately, most subtractive color mixing is complex, and the resulting colors can only be predicted tentatively by applying some rather complex equations. In fact, the equations are so complicated that the use of a computer is essential for their rapid solution. Therefore, we will confine our discussion only to simple subtractive color mixing.

11.5 Subtractive Color Mixing

You can use your own color mixing experience to predict colors and then explain them in terms of subtractive color mixing. This will enable you to be more aware of the difference between pure and impure colors among the pigments you use. For example, mixing red and blue paints produces a violet color. The "red" pigment must absorb all wavelengths except those in the red and violet ranges. (Some reds also reflect a little blue light.) The blue pigment must absorb red, orange, yellow and green, but reflect blue and violet. Thus, red and blue mixed together absorb every wavelength except the violet range, so dark violet light is reflected.

In the same way, red and green produce brown, which is a dark orange yellow. You can account for most color combinations of commercial paints in a similar fashion.

Subtractive color mixing also accounts for the colors seen when light filters are superimposed over one another, as we observed in the preceding chapter. Suppose that we place a yellow filter in front of a white light source. The filter will absorb the blue-violet end of the spectrum, and will transmit red, yellow, orange and green light. If a cyan filter (cyan is the technical term for a certain shade of blue-green) is placed in front of the yellow filter, it will absorb the red-yellow-orange portion of the transmitted light beam, and allow only green light to get through. Thus, superimposition of a yellow filter and a cyan filter, as illustrated in Figure 11.3, will produce the spectral transmittance curve shown in Figure 11.4. Notice that the green light is considerably attenuated (cut down) because much of the incident light has been absorbed by the filters.

If a spectrally pure blue filter had been used in place of the cyan filter, no light would have emerged and the pair of filters would have appeared opaque. This is illustrated in Figure 11.5.

11.6 Subtractive Color Primaries

Because of the impurities in commercial pigments, colorists prefer not to use the color names that artists use, but have adopted more precise terms. Artists speak of red, yellow and blue pigment primaries. Printers, however, who have to be very precise with their color mixing, use the terms magenta, yellow and cyan as their color primaries. Each of these hues is a spectral mixture which reflects two colors. In the subtractive color primary diagram shown in Figure 11.6, each circle reflects the two pure colors it intersects. Thus, magenta reflects wavelengths in the red and blue

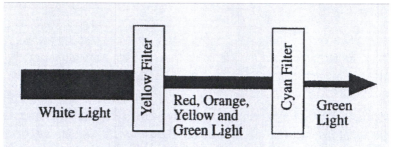

FIGURE 11.3. Absorption of White Light
by Yellow and Cyan Filters

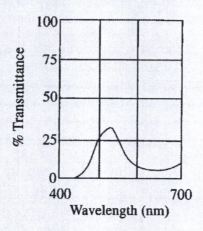

FIGURE 11.4. Transmittance Curve for
Superimposed Yellow and Cyan Filters

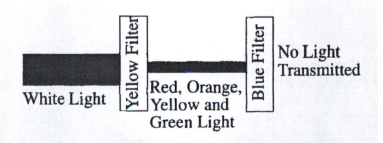

FIGURE 11.5.

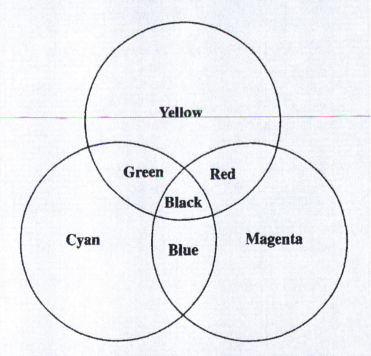

FIGURE 11.6. The Primary Colors for Simple Subtractive Mixing and Their Complements

regions, cyan reflects in the green and blue regions, and yellow reflects in the red and green regions.

The subtractive color primary circles are quite handy to use. The color produced upon mixing any two of the subtractive primaries is shown in the overlap area. Thus, yellow and magenta pigments mix to produce red; cyan and magenta pigments mix to produce blue. All three subtractive color primaries mixed together would absorb the entire visible spectrum and the combination would appear black (center of diagram where all three circles overlap). In actual practice, since no pigment absorbs or reflects ideally, the mixture produces a gray which is very much on the black side.

11.7 Subtractive Complementary Colors

Any two colors opposite each other on the subtractive primary diagram are called complementary colors. Suppose a painter mixes red and cyan pigments together. These are complementary subtractive colors. The red pigment absorbs all of the colors except red; the cyan pigment reflects all of the colors except red. Together, the two pigments absorb all the wavelengths of the visible spectrum: red begins by absorbing all but red, and cyan *completes* the job of absorption by absorbing the red. Hence, the word complementary is used to show that the complete spectrum is absorbed by these two colors when they are mixed. The combination will appear, ideally, black.

Other combinations of pigments or of filters will produce the other colors shown in Figure 11.6. It might be worthwhile to experiment with such filters on your own. The use of filters is far easier and less expensive than mixing of pigments, and similar effects can be achieved.

Subtractive color mixing, in addition to the obvious use of dye and pigment mixing, is also the principle behind color photographic processing, color printing by overprinting, and in producing the color effects seen in the dyeing of transparent plastics.

11.8 Pigment Particles Are Tiny Filters

You may be surprised to learn that most pigments which impart color to paints are not opaque, but are tiny, transparent particles. A paint consists of a suspension of tiny pigment particles in a transparent liquid. The supporting liquid is called the vehicle or medium. The pigment particles are not dissolved in the medium; otherwise, the paint would be transparent because all true solutions are transparent.

The particles of pigment suspended in the medium act just like subtractive filters, as shown in Figure 11.7. Since each pigment particle is transparent, each reflects, refracts and transmits the light striking it. An incident ray will undergo reflection and refraction at each pigment-medium boundary surface, as well as at the medium-air interface. In addition, some of the incident light will be absorbed with each passage though a pigment particle, so multiple passages result in a greatly attenuated reflected ray.

If the ground under the paint is white, it simply reflects back any light that strikes it. Even if the ground is dark, much of the light will be bounced from particle to particle until it finally re-emerges, although less bright, since a dark ground will absorb some of the light. Since the emergent light is disorderly in direction, there is no mirror effect and the paint looks glossy or flat depending on the surface. If the paint consists of yellow and blue pigments, the emergent light will have passed though many particles and each particle will have acted as a tiny filter for its part of the spectrum. As a result, all of the visible spectrum except green will have been absorbed, and the paint will appear green.

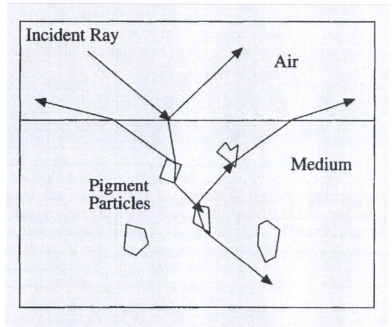

FIGURE 11.7. An Incident ray undergoes multiple reflections and refractions in a pigment. It may be reflected back into the air in the process.

Pigment particle size and the refractive indices of both the pigment and the medium will also affect the interaction of light with the paint. These effects will be discussed in more detail in the chapters on Pigments and Paints.

11.9. Selected Reading

Hope, A.; Walch, M. *The Color Compendium*; Van Nostrand Reinhold: New York, 1990.

Miniexperiment XII

To demonstrate how refraction creates reflection, fill a smooth-sided glass part way with water. Hold the glass a little above your eyes and to one side of your face. Now try to see up through the surface. Insert a pencil into the glass of water. Can you see it above the surface of the water?

Another experiment showing total internal reflection requires a deep dish filled with water, a drinking glass or beaker, and a coin. Place the coin in the bottom of the dish and lower the inverted glass directly down over it making sure that no water enters the glass. Now look for the coin from one side of the glass. Can you see it? Why, or why not? Now allow the glass to fill with water. Can you see the coin now?

Chapter 12

Additive Mixing: The Colors of Lights

CHAPTER 12

ADDITIVE MIXING: THE COLORS OF LIGHTS

12.1 Mixing Colored Lights

Suppose we have set up three stage lights and have covered one with a pure red filter, one with a pure blue filter and one with a pure green filter. Now, let us flash these lights onto a white screen, overlapping them in various combinations, remembering that we are overlapping lights, not filters. A few surprises await us.

To begin with, we are adding lights to one another, so the combined effect when all three lights overlap should be a brighter field than each of the lights viewed alone. This is a very different effect from mixing paints in various combinations. Secondly, when we mix paints together, we expect to see a black or gray area when all three subtractive primaries are added together (Figure 11.6). However, when we overlap all three colored lights in the properly adjusted proportions, we actually get not black, but almost white light. This is because red, green and blue, which are the complementary colors to yellow, magenta and cyan, together complete the spectrum of white light. Thus, they may be termed the additive primary colors; their relationship with one another and their complements, yellow, magenta and cyan, are illustrated in Figure 12.1. Referring to this diagram, we see that where each of two of the circles overlap, we get a combination color. Red plus blue produces magenta; blue plus green yields cyan; green plus red gives yellow. Red plus blue plus green produces white.

12.2 Additive Color Primaries

Additive color mixing can be based on any three colors provided that each falls in a different part of the visible spectrum. The colors red, green and blue are the additive primary colors chosen by physicists for studying mixtures of lights because more colors can be matched by these primaries than any other three. There are other three-color sets that have the same properties as the red-green-blue primaries but, probably for psychological reasons, only red, green and blue are ordinarily considered as primaries. There is also no hard and fast rule about the number of primaries in a color addition system.

If we arrange the colors of the visible spectrum in a circle so that there is a clockwise increase in energy from red to violet (Figure 12.2), we will have formed a color rosette. The number of primary colors and their complements will be determined by how many "slices" one makes out of this circle. Newton, to whom we owe the arrangement of the first color rosette, chose seven primaries plus seven complementaries to yield a rosette with fourteen sections. Frederick E. Ives (1856-1937), another colorist, chose three primaries; Wilhelm Ostwald (1853-1932), a chemist, chose four; Albert Munsell (1858-1918), an artist-colorist, chose five. The choice of a color circle depends mainly upon the use to which it will be put. Figure 12.2 reflects Munsell's choice of primaries (except that Munsell used the term "purple" for "violet"). Theoretically, a color rosette allows for an infinite number of primaries and complementaries.

Interpretation of the color rosette has several useful applications. Complementary-primary pairs can be easily identified. For example, red-cyan, areas opposite one another on the rosette, are complementary. Note that the complement to blue ("artists' blue") is yellow-red. A second application of the color rosette lies in being able to predict a color produced by selective absorption. The

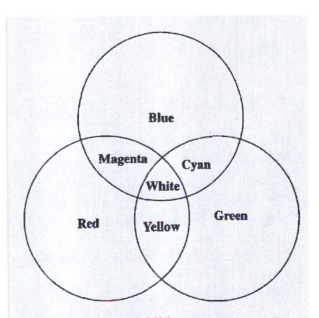

FIGURE 12.1. Additive Primaries and
Their Complements

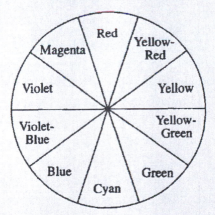

FIGURE 12.2. A Color Rosette

observed color of an object is the complement of the color absorbed. For example, if an object absorbs violet-red light, it will appear green in reflected light. As each pair of complementary colors meets at the center of the rosette, white light is produced.

Interestingly enough, your eyes can directly tell you the complementary color of any color you are examining. To test this statement, draw a filled-in circle on a piece of white paper with a red magic marker or a laboratory marking pencil. Stare at the circle under a bright light for about 40 seconds, and then shift your gaze to another part of the paper. You will see an image of the circle, but in the color complementary to the one you gazed at, namely, cyan. Apparently, by overloading the optical apparatus with a stimulus for red, the red-sensitive apparatus tires and will not react to any red stimulus for a short time period after the stimulus. Since the stimulus coming from the white paper contains red, green and blue stimuli simultaneously, and the eye is failing to react to the red, only the green and blue components of the spectrum will stimulate the eye to see cyan. You can test this hypothesis further by staring at objects of other colors such as green, orange, *etc.*

12.3 Maxwell Color Discs

When you see moving pictures, your brain interprets them as continuous motion even though you are actually seeing 16 still frames per second. This is because of vision persistence: the eye-brain perceptual apparatus seems to hold each image for about 1/10 of a second. Thus, each still frame is overlapped onto the next to give the image of motion.

About a century ago, a Scottish physicist named James Clerk Maxwell (1831-1879) made a device which flashes two or more colors into the eye so rapidly that vision persistence caused them to blend additively. Maxwell painted two discs each a different color. Then he bisected each disc to the center and overlapped them via the bisecting slits. The combined discs were placed on a wheel and spun rapidly. By altering the proportions of each color, various additive hues could be observed.

12.4 Spatial Averaging

Additive color mixing also takes place with adjacent objects when they are viewed from a sufficient distance. This is called spatial averaging. This is the principle used in both color printing and in producing the image of a color television set. A color television screen is coated with a regular array of three different phosphors which glow red, green and blue respectively when energized by an electron beam. White is created in the tube when the beams energize each phosphor dot equally; the absence of an energy beam causes them to appear black. The complete "living color" spectrum seen on TV is really red-green-blue living color. When electron beams energize a red dot and a green dot adjacent to each other, then the stimulus reaching our eye is red-green. Since the dots are so close to one another, our eye is incapable of separating them, or resolving them, into two separate dots and responds to them as a single red-green = yellow dot. If you examine a "yellow" area on a TV screen with a good magnifying glass, you may be able to resolve (seem to pull apart) the red and green dots.

Most printed color illustrations also use dots of the primary additive colors. To do this, the photoengraver starts by placing a fine-mesh wire screen in front of the camera. Then, a yellow filter is placed over the lens and the picture is photographed to copy all of the yellow in the picture. This produces a negative with yellow dots, and from this a printing plate is made that is inked with yellow. Similarly, cyan and magenta plates are made, but only after rotating the wire screen about 15° between each photograph of the color negatives in order to prevent superimposition of dots. Thus visual averaging takes place along with some subtractive color mixing, to produce all of the colors of the original. If the inks were pure colors, black would be produced in the picture by overlap of all three colors. However, since the inks are impure, a black plate is added to the other three. Modern presses print all four impressions so quickly that they can do about 12,000 impressions per hour. A good magnifying glass will help you to pick out the differently colored dots in a color print.

12.5 The Impressionists and Additive Color Mixing

During the late Nineteenth Century, a greater variety of bright pigments became available to artists. Simultaneously, a new interest in the scientific study of color began to develop. The culmination of these two developments came in the innovative work of the Impressionistic School. Whereas the early Renaissance artists used "descriptive" color in order to create a lasting, timeless feeling, the Impressionists sought to capture the fleeting quality of each daily moment. Much of their work was done outdoors *(plein-air* painting) since they could observe the effect of the sunlight at different times of day. Claude Monet (1840-1926) painted a lily pond over and over again, and it never appeared to be the same in any two paintings.

To capture the changing, shimmering quality of light, the Impressionists began to employ short strokes of pure colors which spatially averaged to clear, brilliant hues. This technique overcame the problem of the "muddiness" caused by mixing two pigments together. At first, they used complementary colors in the shadows. Later, they used them for the entire painting. Not only were the colors more brilliant, but a feeling of life and movement was added by the tiny dabs of color.

The ultimate extension of this technique came when some painters began to paint with tiny dots of color. This technique, called Pointillism required extraordinary patience and a thorough knowledge of additive color mixing. Georges Seurat (1859-1891) is one of the prime examples of an artist who employed this painstaking technique. A typical painting by Seurat took about two years to complete.

12.6. Selected Readings

Bomford, David, *et al. Art in the Making: Impressionism*; The National Gallery: London, 1990.

Falk, David, *et al. Seeing the* Light; Harper and Row: New York, 1986.

Ratliff, F. *Paul Signac and Color in Neo-Impressionism*; Rockefeller University Press: New York, 1992.

Miniexperiment XIII

Try additive color mixing by taping cellophane or gelatin filters to flashlights or mounting them on slide projectors. Based upon the colors produced, estimate the spectral transmittance curves for your filters.

Make Maxwell discs and spin them on a spinner toy top, an electric drill, or simply mount them on a stick and spin. Cut-outs of colored construction paper may be used, but you will probably find these colors very impure. Opaque poster paints work fairly well, but any colorants, including crayons, may be used. Try various combinations of two or more colors. An infinite variety is possible. Try mixing colors with white and then with black. Compare rapid and slow spinning. You may want to use an old record-player turntable for this purpose (if available). Test the colors in your pigment tubes by painting them out on Maxwell discs; this will give you information regarding their purity. Try mixing a cyan and a red disc. Since cyans are mixtures of green and blue, addition of red should provide a visual stimulus of white. In actual practice, you will probably see a light gray. Increasing the area of the red disc will yield a pink. Spinning a red and a green disk should give a yellow color (unlike the muddy brown of subtractive mixing). You may have to alter the proportions (disc areas) of the pigments to get a good yellow.

Use the above to compare additive and subtractive mixing. Cut out a disc of cardboard and trace it onto three parts of a white sheet. On the sheet, paint one disc yellow and one blue. Next, blend paint exactly half yellow and half blue, mix well, and paint onto the third disc. That disc shows subtractive mixing. Paint the cardboard disc half yellow and half blue on one side. Spin to get additive mixing. Compare. Try other complementary colors.

Part III

The Chemistry of Color

CHAPTER 13

THE SHORTHAND OF CHEMISTRY

13.1 Chemical Symbols

Chemists have their own form of shorthand which enables them to present ideas quickly and efficiently. The use of symbols for the chemical elements, of formulas showing which elements are linked together, and in what ratio, and of equations which symbolize exchanges in atomic partnerships are probably the most widely used forms of this shorthand. There is nothing mysterious or exotic about it. It is merely a convenience which we shall find very useful in this text.

As mentioned previously in Chapter 7, all of the known elements have symbols assigned to them which contain one or two letters; the first letter is always capitalized and the second is always in lower case. We also learned that each element had an average atomic mass based upon carbon-12. Table 13.1 lists the elements in alphabetical order and includes the symbol, atomic number and atomic mass of each. (Masses designated with † are for the most commonly available long-lived isotope; masses with * are for the most stable or best-known isotope.) As we use these symbols, you will become familiar with them; your instructor may suggest a list for you to memorize.

Each symbol refers to a particular element and may refer to (a) one atom of the element; (b) one mole of the element (6.02×10^{23} atoms of the element) or (c) the element in general, depending on the context of the sentence or the placement of the symbol in the chemical equation.

TABLE 13.1

ALPHABETICAL LIST OF THE ELEMENTS

Element	Symbol	Atomic Number	Atomic Mass
Actinium	Ac	89	227.0278†
Aluminum	Al	13	26.98154
Americium	Am	95	(243)*
Antimony	Sb	51	121.75
Argon	Ar	18	39.948
Arsenic	As	33	74.9216
Astatine	At	85	(210)*
Barium	Ba	56	137.34
Berkelium	Bk	97	(247)*
Beryllium	Be	4	9.01218
Bismuth	Bi	83	208.9804
Bohrium	Bh	107	(270)*
Boron	B	5	10.81
Bromine	Br	35	79.904
Cadmium	Cd	48	112.40
Calcium	Ca	20	40.08
Californium	Cf	98	(251)*
Carbon	C	6	12.011
Cerium	Ce	58	140.12
Cesium	Cs	55	132.9054
Chlorine	Cl	17	35.453
Chromium	Cr	24	51.996
Cobalt	Co	27	58.9332
Copernicium	Cn	112	(285)*
Copper	Cu	29	63.546
Curium	Cm	96	(247)*
Darstadtium	Ds	110	(281)*
Dubnium	Db	105	(268)*
Dysprosium	Dy	66	162.50
Einsteinium	Es	99	(252)*
Erbium	Er	68	167.26
Europium	Eu	63	151.96
Fermium	Fm	100	(257)*
Flerovium	Fl	114	(289)*
Fluorine	F	9	18.998403
Francium	Fr	87	(223)*
Gadolinium	Gd	64	157.25

TABLE 13.1 (continued)

Gallium	Ga	31	69.72
Germanium	Ge	32	72.59
Gold	Au	79	196.9665
Hafnium	Hf	72	178.49
Hassium	Hs	108	(269)*
Helium	He	2	4.00260
Holmium	Ho	67	164.9304
Hydrogen	H	1	1.0079
Indium	In	49	114.82
Iodine	I	53	126.9045
Iridium	Ir	77	192.22
Iron	Fe	26	55.847
Krypton	Kr	36	83.80
Lanthanum	La	57	138.9055
Lawrencium	Lr	103	(262)*
Lead	Pb	82	207.2
Lithium	Li	3	6.941
Livermorium	Lv	116	(293)*
Lutetium	Lu	71	174.967
Magnesium	Mg	12	24.305
Manganese	Mn	25	54.9380
Meitnerium	Mt	109	(278)*
Mendelevium	Md	101	(258)*
Mercury	Hg	80	200.59
Molybdenum	Mo	42	95.94
Neodymium	Nd	60	144.24
Neon	Ne	10	20.179
Neptunium	Np	93	237.0482†
Nickel	Ni	28	58.71
Niobium	Nb	41	92.9064
Nitrogen	N	7	14.0067
Nobelium	No	102	(259)*
Osmium	Os	76	190.2
Oxygen	O	8	15.9994
Palladium	Pd	46	106.42
Phosphorus	P	15	30.97376
Platinum	Pt	78	195.09
Plutonium	Pu	94	(244)*
Polonium	Po	84	(209)*
Potassium	K	19	39.0983
Praeseodymium	Pr	59	140.9077

TABLE 13.1 (continued)

Promethium	Pm	61	(145)*
Protactinium	Pa	91	231.0359†
Radium	Ra	88	226.0254†
Radon	Rn	86	(222)*
Rhenium	Re	75	186.207
Rhodium	Rh	45	102.9055
Roentgenium	Rg	111	(280)*
Rubidium	Rb	37	85.4678
Ruthenium	Ru	44	101.07
Rutherfordium	Rf	104	(267)*
Samarium	Sm	62	150.36
Scandium	Sc	21	44.9559
Seaborgium	Sg	106	(269)*
Selenium	Se	34	78.96
Silicon	Si	14	28.0855
Silver	Ag	47	107.868
Sodium	Na	11	22.98977
Strontium	Sr	38	87.62
Sulfur	S	16	32.06
Tantalum	Ta	73	180.9479
Technetium	Tc	43	98.9062†
Tellurium	Te	52	127.60
Terbium	Tb	65	158.9254
Thallium	Tl	81	204.37
Thorium	Th	90	232.0381†
Thulium	Tm	69	168.9342
Tin	Sn	50	118.69
Titanium	Ti	22	47.90
Tungsten	W	74	183.85
Ununoctium	Uuo	118	(294)*
Ununpentium	Uup	115	(288)*
Ununseptium	Uus	117	(294)*
Ununtrium	Uut	113	(286)*
Uranium	U	92	238.0289
Vanadium	V	23	50.9415
Xenon	Xe	54	131.29
Ytterbium	Yb	70	173.04
Yttrium	Y	39	88.9059
Zinc	Zn	30	65.38
Zirconium	Zr	40	91.22

You may have noticed that some elements with atomic numbers above 100 have three-letter symbols. These are provisional symbols assigned by the International Union of Pure and Applied Chemistry (IUPAC) for those elements that have been claimed to have been discovered (synthesized) by two different groups working in the United States and in the former Soviet Union. Their names and symbols correspond to their atomic numbers (115 is "un" for one and "pent" for five - ununpentium) and corresponding three letter symbols. For more information, please see Orna, M.V., "On Naming the Elements With Atomic Number Greater Than 100," *Journal of Chemical Education* **59,** p. 123 (1982).

13.2 Chemical Formulas

When atoms unite to form a stable combination, the combination is called a chemical compound. The ratio of the number of atoms of each element present in the compound is shown by the formula. The symbols in the formula identify the element, and the subscript shows the number of atoms. For example, the formula for water is H_2O, indicating that there are two atoms of hydrogen for each atom of oxygen in the compound. The subscript 1 is conventionally omitted, and you are expected to know that whenever no subscript is used, there is only one atom of that element in the formula. The formula of sulfuric acid is H_2SO_4. There are two atoms of hydrogen, one atom of sulfur and four atoms of oxygen per unit of sulfuric acid. Other formulas of compounds you will encounter later are a form of iron oxide, Fe_2O_3, calcium carbonate, $CaCO_3$, silver bromide, $AgBr$, and oleic acid, $C_{17}H_{33}COOH$. Sometimes formulas contain parentheses such as the formula of calcium phosphate, $Ca_3(PO_4)_2$. In this case, the subscript 2 applies to everything within the parentheses, so that a unit of calcium phosphate contains 3 atoms of calcium, 2 atoms of phosphorus and 8 atoms of oxygen. Count the number of atoms in the following formulas: $Mg(HCO_3)_2$, $(NH_4)_2SO_4$, $Na_2S_2O_3$.

As you might expect, the identity of a substance depends upon the atoms that make it up, and upon the number and arrangement of these atoms in the material in question. These factors determine the chemical and physical properties of the substance. If you change any of the atoms or even rearrange atoms in a compound, you will no longer have the same compound, but a new one with different properties. For example, the two compounds that carbon forms with oxygen are CO, carbon monoxide, and CO_2, carbon dioxide. They differ in the number of oxygen atoms associated with the carbon, but this difference has a tremendous effect on their physical and chemical properties. While both are colorless, odorless gases, CO is slightly less dense than air and CO_2 is about 1.5 times more dense than air. CO is a poisonous gas, and CO_2 is a gas necessary to the life of green plants, and thus to our own lives. Another example is the difference between sodium sulfate, Na_2SO_4, a compound sometimes called Glauber's salt, and sodium thiosulfate, $Na_2S_2O_3$, a compound in which a sulfur has replaced an oxygen in Glauber's salt. The former compound is used in the manufacture of ceramic glazes, dyes and soaps; the latter, sometimes called sodium hyposulfate, is the famous "hypo" which is used as a fixing agent in photographic processing.

In these two examples, we have seen how chemical properties can be affected by adding or replacing an atom in a compound to form a new compound. We will also examine the changes that can be effected by simply rearranging atoms. However, we will have to wait until the device of the structural formula is introduced in the chapter on organic compounds.

Since each chemical compound has a definite formula, you might surmise that there are some rules that govern the formation of compounds, and therefore, the writing of formulas as well. These "rules" have their origin in the structures of the atoms themselves, and so we will have to wait until we have examined atomic structure more closely in order to learn why compounds have definite formulas.

13.3 Chemical Equations

In the early days of chemistry, the formula of a compound was determined by actual chemical analysis. To find out the composition of limestone, for example, a chemist would first test to see which elements were present, and then how much of each element. At first, limestone was found to consist of three elements, calcium, carbon and oxygen. Further analysis showed that for approximately 40 grams of calcium present, 12 grams of carbon and 48 grams of oxygen were present. Since each atom of calcium has a 40/12 weight ratio to each atom of oxygen (See Table 13.1 for the relative atomic masses of these elements and round off to whole numbers.), we conclude that the ratio of calcium to carbon to oxygen in limestone is 1 to 1 to 3. Hence the formula is $CaCO_3$.

It has long been known that limestone deteriorates in the presence of acids. Limestone buildings and monuments suffer from this unfortunate circumstance due to the acid rains characteristic of highly industrialized areas. It is easy to observe this deterioration in the kitchen or in the laboratory. If one pours an acid such as hydrochloric acid (HCl) or even a weak acid such as vinegar (acetic acid, CH_3COOH) over small pieces of limestone, marble chips or chalk (all are forms of the same compound), effervescence occurs and the limestone partially dissolves. If enough acid is added, all of the limestone will disappear with the vigorous evolution of a colorless, odorless gas. If the gas were collected, and if it were analyzed both qualitatively (to find the kind of each element present) and quantitatively (to find out how much of each element is present), the formula of the gas would turn out to be CO_2. The gas is carbon dioxide, the same gas that bubbles out of a carbonated beverage as it is being poured. Further analysis of the remaining solution would reveal that two additional compounds were formed if HCl were the acid used, calcium chloride, with the formula $CaCl_2$, and water, H_2O.

A convenient way to describe the chemical reaction which takes place when limestone deteriorates by the action of hydrochloric acid is the following sentence: Calcium carbonate (limestone) plus hydrochloric acid yields upon analysis calcium chloride, water and carbon dioxide. In symbolic form:

$$CaCO_3 + HCl \rightarrow CaCl_2 + H_2O + CO_2 \qquad [13.1]$$

The arrow in the above statement is always read as "yields." Since the limestone is a solid, the water is a liquid, and the carbon dioxide is a gas, and the hydrochloric acid and the calcium chloride are both in aqueous (water) solution, a more refined version of the statement is:

$$CaCO_3(s) + HCl(aq) \rightarrow CaCl_2(aq) + H_2O(l)\ CO_2(g) \qquad [13.2]$$

This is now a very neat and compact way to represent just what happens when limestone and hydrochloric acid react, provided we know the chemical formulas of the reactants, *i.e.,* the substances to the left of the arrow, and of the products, *i.e.,* the substances to the right of the arrow.

However, chemists are very unhappy with the above statement as it stands because it violates the most basic law of chemistry, which is that in any change that takes place in nature, matter is conserved. In other words, matter can neither be created nor destroyed, but only rearranged. Note that in the above statement, only one atom each of hydrogen and chlorine appear among the reactants, but two atoms of each appear among the products. This is clearly impossible. We must always end up with the same numbers and kinds of atoms before and after a change of this sort. Otherwise, I end up "creating" hydrogen atoms and chlorine atoms out of nothing by doing this reaction.

We can be helped out of this dilemma by recognizing that we are never dealing with a single unit of $CaCO_3$ or a single unit of HCl. Just a single gram of limestone contains about 6×10^{21} units of $CaCO_3$, and a single gram of pure HCl contains about 1.64×10^{22} units of HCl. Thus, when $CaCO_3$ reacts with HCl, they are not constrained to react 1 to 1. As a matter of fact, you can straighten out our problematic statement above by assuming that for every 1 unit of $CaCO_3$, 2 units of HCl react. Writing this assumption in the statement gives:

$$CaCO_3(s) + 2HCl(aq) \rightarrow CaCl_{2}(aq) + H_2O(l) + O_2(g) \qquad [13.3]$$

We have cleverly introduced a "2" in front of the HCl to indicate that two units of HCl react. The "2" is called a coefficient, and it applies to the complete formula that follows it, *i.e.,* to the HCl. Thus the insertion of the coefficient takes care of our creation of matter dilemma because now two atoms each of hydrogen and chlorine appear on both sides of the statement. In fact, now that the same number of each type of atom appears on both sides of the statement, in reality we are dealing with an equality. Therefore, this complete statement is now called a chemical equation. Our introduction of the coefficient balanced the equation, and strictly speaking, the statement is not an equation unless or until it is balanced. Additional examples of balanced chemical equations are:

$$2CO(g) + O_2(g) \rightarrow 2CO_2(g) \qquad [13.4]$$
$$CaCO_3(s) + 2CH_3COOH(aq) \rightarrow (CH3COO)_2Ca(aq) + H_2O(l) + CO_2(g) \quad [13.5]$$
$$Fe_2O_3(s) + 3H_2(g) \rightarrow 2Fe(s) + 3H_2O(g) \qquad [13.6]$$

Verify that each of these equations is balanced. Note that in balancing a chemical equation, only the coefficients, not the subscripts within each formula, can be changed. Otherwise, you change the identity of the substance reacting. Note also that the coefficient of "1" is understood by convention.

You may be wondering how a chemist knows what numbers to insert as coefficients. There are several formal ways to go about balancing an equation, but simple equations are readily balanced by a trial-and-error method. With a little practice, trying out numbers until a combination is found which balances the equation becomes quite easy to do. Try your hand at balancing the following:

$$H_2SO_4(aq) + NaOH(aq) \rightarrow Na_2SO_4(aq) + H_2O(l) \qquad [13.7]$$

$$CaCO_3(s) + H_2SO_4(aq) \rightarrow CaSO_4(aq) + H_2O(l) + CO_2(g) \qquad [13.8]$$

$$Ca(OH)_2(s) + H_3PO_4(aq) \rightarrow Ca_3(PO_4)_2(aq) + H_2O(l) \qquad [13.9]$$

$$FeO(s) + O_2(g) \rightarrow Fe_2O_3(s) \qquad [13.10]$$

You may find that it is quite easy to balance the above equations, but not so easy to balance the following:

$$Cu(s) + HNO_3(aq) \rightarrow Cu(NO_3)_2(aq) + NO(g) + H_2O(l) \qquad [13.11]$$

Formal methods are required to balance many chemical equations, but we will not have occasion to introduce these methods until we reach the chapter on photography.

13.4 The Significance of the Chemical Equation

The balanced chemical equation contains a wealth of information of a quantitative as well as a qualitative nature. Not only does it tell us what products to expect from what reactants, but also the numerical ratios in which they react or are produced. For example, equation [13.4] tells us that two units of gaseous carbon monoxide react with one unit of gaseous oxygen to produce two units of gaseous carbon dioxide.

From Table 13.1, the relative atomic masses of carbon and oxygen are 12 and 16, respectively (rounded off to the nearest whole numbers). Thus, the relative masses of CO, O_2 and CO_2 are 28, 32 and 44. Since two units of CO, $i.e.,$ 56 relative mass units, react with 1 unit of O_2, $i.e.,$ 32 relative mass units, we know that in this particular reaction, we will always need a mass ratio of 56/32 for CO/O_2 for these two substances to react with one another completely. If extra CO or O_2 is present, the extra units will not react. Put another way, if I take 56 grams of CO, I know that I will need exactly 32 grams of O_2* for reaction. If I add 34 grams of O_2 instead, two grams of O_2 will remain unreacted.

It would take a good deal more time to teach you how to do these calculations on your own. However, the point is this: the chemical equation gives information regarding exactly how much of each chemical will react in a particular reaction in terms of mass ratios. This means that the chemist can set up a reaction so that there are no "leftover" reactants at the end of the reaction. This is particularly important in dealing with the formation of compounds which would be less beautiful if, at the end of the reaction, some of the unreacted starting material were mixed in. For example, you probably know that ceramic artists must weigh out each ingredient for a glaze recipe very carefully. This is because all of the ingredients react with one another in a definite mass ratio

---

*O_2 indicates two atoms of oxygen bound together. It is the formula for the element oxygen as it exists at ordinary temperatures. Since a chemical equation deals with reality, that is, with elements and compounds as they exist, we must use the formula O_2, and not O or O_3. Other elements that exist at ordinary temperatures in this double form, called "diatomic," are H_2, N_2, Cl_2, Br_2, F_2 and I_2. However, they are only diatomic as elements. They may exist in other combinations in compounds, $e.g.,$ HCl or $NaIO_3$.

that has been calculated beforehand. If an extra amount of one of the ingredients is added "for good measure," it will remain unreacted in the firing process and could possibly cause the glaze to develop cracks, or it might simply remain dispersed as an unsightly crystalline material throughout the finished glaze.

Before we can use some of the new knowledge from this chapter in order to formulate our own compounds, we have to learn something about the relationships of the elements among themselves and the structures of their atoms. Atomic structure, in the end, determines these relationships, and how and how many atoms will interact with one another. The next two chapters, "The Periodic Table" and "Electron Configurations" will provide this information.

13.5 Suggested Readings

Aldersey-Williams, H. *Periodic Tales: The Curious Lives of the Elements*. Penguin Books: London, 2011

Kean, Sam. *The Disappearing Spoon (and other true tales of madness, love and the history of the world from The Periodic Table of the Elements)*. Back Bay Books: Costa Mesa, CA, 2010. Warning: The strength of this book is the background information on the people involved; tread carefully when the author discusses chemistry, or use it as a rich source of assessment questions of the "defend or refute" variety.

CHAPTER 14
THE PERIODIC TABLE

14.1 The Weighting Game

In the century or so prior to 1869, chemists were in a flurry of activity over the discovery of new element after new element. While many elements were isolated by purely chemical reactions, a breakthrough in the very early nineteenth century, the construction of the first electric battery, led Sir Humphry Davy (1778-1829) to isolate new elements by electrical means. By 1869, some 60-odd elements had either been isolated, or their existence had been predicted. In addition, the relative atomic masses and the chemical and physical properties of most of these elements were known.

The accumulation of all this new data, and the prospects of accumulating even more, caused two chemists, Lothar Meyer (1830-1895) in Germany and Dmitri Mendeleev (1834-1907) in Russia, to attempt to organize these data in some way. Their basic assumption was that nature must be orderly if similarities in properties of many of the newly discovered elements were observed again and again. Since one of the chief distinguishing characteristics among the elements was their relative atomic masses, Mendeleev, to whom the credit finally went for this attempt at organization, decided to arrange the then-known elements in order of increasing atomic mass. The year was 1869. Table 14.1 is a replica of the list Mendeleev must have worked from. The elements known in 1869 are listed in order of their then-known atomic masses, and their modern atomic masses are given in parentheses. Since the symbols of several of the elements have changed since Mendeleev's time, the symbols he used for these elements are given in parentheses.

TABLE 14.1:
THE KNOWN ELEMENTS IN 1869

Element (Symbol)	Relative Atomic Mass	Date of Discovery	Element (Symbol)	Relative Atomic Mass	Date of Discovery
H	1	1766	Br	80	1826
He	4	1868	Rb	85.4	1861
Li	7	1817	Sr	87.4	1808
Be	9.4	1798	Zr	90 (91)	1789
B	11	1808	Ce	92 (140)	1803
C	12	Antiquity	La	94 (139)	1839
N	14	1772	Nb	94 (93)	1801
O	16	1774	Nd (Di)	95 (144)	1841
F	19	Iso. 1886	Mo	96	1778
Na	23	1807	Rh	104 (103)	1803
Mg	24	1808	Ru	104 (101)	1827
Al	27.4	1827	Pd (Pl)	106.5	1803
Si	28	1800	Ag	108	Antiquity
P	31	1669	Cd	112	1817
S	32	Antiquity	U (Ur)	116 (238)	1789
Cl	35.5	1774	Sn	118 (119)	Antiquity
K	39	1807	Th	118 (232)	1828
Ca	40	1808	Sb	122	Antiquity
Ti	50 (48)	1791	I	127	1811
V	51	1801	Te	128	1782
Cr	52	1797	Cs	138 (133)	1860
Mn	55	1774	Ba	137	1774
Fe	56	Antiquity	Ta	182 (181)	1802
Er	56 (167)	1842	W	186 (184)	1783
Ni	59	Antiquity	Pt	197 (195)	1735
Co	59	1735	Ir	198 (192)	1803
Y (Yt)	60 (89)	1794	Os	199 (190)	1803
Cu	63.4	Antiquity	Hg	200 (201)	Antiquity
Zn	65.2	Antiquity	Au	197	Antiquity
As	75	13th C.	Bi	210 (209)	1753
In	76 (115)	1863	Tl	204	1861
Se	79.4	1817	Pb	207	Antiquity

14.2 Mendeleev's Feat

Arranging data in a table such as Table 14.1 is not difficult. However, Mendeleev went much further, and in so doing, paved the way for many discoveries in the ensuing decades.

First of all, Mendeleev recognized that many of the elements exhibited similar chemical properties. He noticed that certain non-metals with the general formula X_2 reacted with sodium in the following manner:

$$2Na + X_2 \rightarrow 2NaX \qquad [14.1]$$

Since these nonmetals happened to be fluorine, F_2, chlorine, Cl_2, bromine, Br_2, and iodine, I_2, four specific chemical equations could be written by simply substituting the proper chemical symbol for each element in place of the X. For example, chlorine reacts thus:

$$2Na + Cl_2 \rightarrow 2NaCl \qquad [14.2]$$

The product of the reaction in equation [14.2] is NaCl, ordinary table salt, which is a white crystalline powder which dissolves in water to yield a colorless solution. Similarly, F_2, Br_2 and I_2 formed NaF, NaBr and NaI, all with similar properties to NaCl.

On the other hand, Mendeleev also noticed that certain active metals such as Li, K, Rb and Cs behaved very much like sodium, Na, when they reacted with nonmetals. The salts LiCl, KCl, RbCl and CsCl exhibited properties similar to NaCl and could be formed by the same general reaction:

$$2M + Cl_2 \longrightarrow 2MCl \qquad [14.3]$$

where M is the general symbol for each of these metals. Similarly, the elements Be, Mg, Ca, Ba and Sr all reacted with Cl_2 in this manner

$$M + Cl_2 \rightarrow MCl_2 \qquad [14.4]$$

On the basis of these observations, Mendeleev decided to rearrange the data in Table 14.1 by laying out the list of elements in vertical columns, but by breaking the sequence of the list in such a way that the elements exhibiting similar chemical properties were lined up in horizontal rows. At the same time, he tried to preserve the order of increasing atomic mass. Let us make an attempt to do what Mendeleev did with his data and see what conclusions we draw.

As a chemist, I can look through the list in Table 14.1 and find all of the elements that react with chlorine according to equation [14.2]. They are Li, Na, K, Cu, Rb, Ag and Cs. Let us make a series of vertical columns with each of these elements at the top of each column, and arrange the elements under them in strict order of increasing atomic mass:

Li	Na	K	Cu	Rb	Ag	Cs
Be	Mg	Ca	Zn	Sr	Cd	Ba
B	Al	Ti	As	Zr	U	Ta
C	Si	V	In	Ce	Sn	W
N	P	Cr	Se	La	Th	Pt
O	S	Mn	Br	Nb	Sb	Ir
F	Cl	Fe		Nd	I	Os
		Co		Mo	Te	Hg
		Y		Rh		Au
				Ru		Bi
				Pd		Tl
						Pb

At first glance, this arrangement looks rather promising. The elements of the first horizontal row all react according to equation [14.2] and the elements of the second horizontal row all react according to equation [14.4]. However, there any resemblance between these elements ceases. Li, Na, K, Rb and Cs are all very active metals and even react violently with relatively inert compounds such as water. Cu and Ag, on the other hand, are relatively unreactive. Similar discrepancies occur in the second row. The following rows show such great irregularities that the utility of this arrangement may be called into serious question. B and Al, C and Si, N and P, O and S, and F and Cl all show similar chemical properties, but the other members of their respective rows are quite unlike them. Br has similar properties to F and Cl, but appears in the row above them. At this point, Mendeleev reasoned that if the table worked for the first couple of rows, however limpingly, the reason why it ceased to work later on is that some elements were probably missing from his list. After all, there were large gaps between a rather regular progression of atomic masses, so why not try to leave room for some missing elements? This was the great breakthrough: if one could leave room for some yet-to-be-discovered elements which should fall into line with the elements on either side of them whose properties were already known, then the properties of the missing elements should be predictable! Furthermore, Mendeleev also called into question the accuracy of some of the reported atomic masses of the known elements, since different masses would shift the position of the elements in his arrangement. A glance at Table 14.1 shows that Mendeleev had good reason for his skepticism; some of the atomic masses were in error by a factor of almost 100%.

Working on these modifying assumptions, Mendeleev eventually developed the table given here as Table 14.2, an arrangement which was published in the *Journal of the Russian Chemical Society* in 1869. Several features of this table should be pointed out. First of all, Mendeleev did not know what to do with helium, a totally unreactive gas which was known to be present in the sun. So he simply left it out. (It was not until the turn of the 20th century that helium's counterparts, neon, argon and krypton, were discovered.) Secondly, he left room for elements which were as yet undiscovered (at atomic masses 45, 68 and 70), and

TABLE 14.2
MENDELEEV'S PERIODIC TABLE AS PUBLISHED IN
THE JOURNAL OF THE RUSSIAN CHEMICAL SOCIETY
(1869)

			Ti=50	Zr=90	? =180
			V=51	Nb=94	Ta=82
			Cr=52	Mo=96	W=186
			Mn=55	Rh=104	P =197
			Fe=56	Ru=104	Ir=198
			Ni=Co=59	Pl=107	Os=199
H=1			Cu=63.4	Ag=108	Hg=200
	Be=9.4	Mg=24	Zn=65.2	Cd=112	
	B=11	Al=27.4	? =68	Ur=116	Au=197?
	C=12	Si=28	? =70	Sn=118	
	N=14	P=31	As=75	Sb=122	Bi=210
	O=16	S=32	Se=79.4	Te=128?	
	F=19	Cl=35.5	Br=80	I=127	
Li=7	Na=23	K=39	Rb=85.4	Cs=138	Tl=204
		Ca=40	Sr=87.4	Ba=137	Pb=207?
		? =45	Ce=92		
		?Er=56	La=94		
		?Yt=60	Di=95		
		?In=75.6	Th=118?		

predicted that these elements should have properties similar to their horizontal counterparts, so he named them respectively *ekaboron, ekaaluminum* and *ekasilicon.* Mendeleev lived to see the discovery of these three *"eka"* elements, known today as scandium, gallium and germanium, along with a host of other elements which may never have been sought were it not for the predictive properties of his wonderful table. Since his table was organized on the basis of recurring, or periodic, properties of the elements, it was eventually called "The Periodic Table." It is one of the most useful tabulations ever devised by the human mind. For his accomplishment, so the story goes, Mendeleev missed out on the 1906 Nobel Prize in chemistry by just one vote. He died, no doubt of other causes, the following year. However, his place in the history of the discovery of the elements was duly recognized when, in 1955, the newly discovered Element 101 was named Mendelevium (Md) in his honor.

14.3 The Modern Periodic Table

When Mendeleev tried to put his table into its final form, he encountered several serious problems in trying to adhere to the arrangement of increasing atomic mass. Chief among these difficulties was the placement of iodine, I, and tellurium, Te. In order of increasing atomic mass, I should precede Te in the table. However, I most closely resembles Br, Cl and F in chemical properties, and Te most closely resembles Se, S and O. The logical thing to do was to switch their positions, but this switch destroyed the mass sequence of the elements, a fact that Mendeleev recorded with a question mark. This little question was probably the first clue to the idea that perhaps mass was **not** the factor that either differentiated the elements from one another or conferred upon the elements their particular properties.

Another problem, as we observed previously, involved the place of helium, the only known inert gas, in the table. It seemed to have no place at all. Other difficulties involved the placement of zinc and cadmium in the same row with beryllium and magnesium since beryllium and magnesium had very similar properties to calcium, strontium and barium. Another difficulty involved the placement of thallium in the same row with the series of very active metals headed by lithium.

However, there were certainly some very good things that could be said about the table. As new elements were discovered, many of them fit into the blanks left by Mendeleev since their predicted properties matched those of the elements beside them in the row. But perhaps the greatest triumph for the Periodic Table came at the very end of the nineteenth century when Sir William Ramsay (1852-1916) and Morris W. Travers (1872-1961) discovered a whole family of relatively inert gases that matched the properties of helium in every respect. These gases, neon, argon, krypton, and later xenon, fit very neatly next to helium in the Table. However, more "trouble" came up when the atomic mass of argon was determined and it was found to exceed that of potassium by almost one atomic mass unit. If the "weighting game" were still to apply, argon should come after potassium in the Table, but then it would be in the wrong row with respect to chemical properties. Therefore, it had to be accorded the same exception as the iodine-tellurium pair had been given. But the vague uneasiness regarding the importance of atomic masses was reinforced by this exception.

It is easy for us moderns to look back on all this and say "Of course! Atomic number is the important quantity, not atomic mass." But we must keep in mind that in 1900, chemists as yet knew nothing about the existence of subatomic particles, and it was to be another decade or two before the elucidation of atomic structure allowed for a re-structuring of the Periodic Table and a recognition of the fact that the Table itself was a shorthand clue to the actual internal structure of the elements!

The arrangement of the modern Periodic Table is given in Figure 14.1. The Table went through many forms before reaching this version, and periodically (no pun intended), chemists come up with suggestions for a different arrangement. This table is different, in several obvious respects, from Mendeleev's table. First, the elements are arranged horizontally in order of increasing atomic number. Secondly, a break in the sequence comes after Ca in order to accommodate the elements Sc through Zn, elements which have no immediate precedents. The sequence resumes with Ga, which has properties quite similar to Al above it. Thirdly, He and its family of so-called "noble gases" are placed on the extreme right. Fourthly, those problematic elements, Nd, Tb and Er, are found in a separate row which should be inserted between Ba and Hf in the table, thus necessitating another break. However, this second break would cause the table to become physically too wide, and therefore inconvenient. So elements 57-71, commonly called the lanthanides because the first member of the sequence is lanthanum, La, and elements 80-103, called the actinides, are given a separate place at the bottom of the table. The step-like heavy line that begins between B and Al is the line that separates the metals from the nonmetals. Note that there are many more metals than non-metals. However, those elements adjacent to the heavy line have often been found to exhibit the properties of both metals and nonmetals, depending upon the circumstances, and are sometimes called metalloids.

There are numerous advantages to the modern Periodic Table, some of which will not become evident until we deal with atomic structure and chemical bonding. However, just an

186

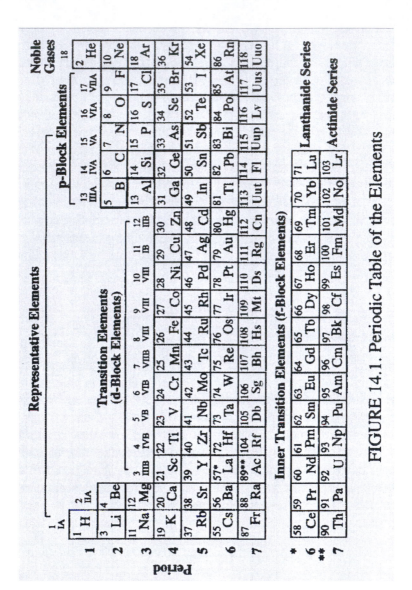

FIGURE 14.1. Periodic Table of the Elements

examination of observed trends in physical and chemical properties justifies its widespread use. First of all, in moving across the table from left to right in any horizontal row, the first noticeable change is the trend from metallic to nonmetallic properties. Metals are materials that conduct electricity, can be hammered into sheets, can be drawn out into wires, and exhibit a metallic luster. On the other hand, nonmetals do not conduct electricity, and are brittle and non-lustrous. The trend is not so obvious in the first horizontal row going from H to He, principally because H itself is a nonmetal. Let us look at the fourth horizontal row, which contains 18 members, in order to see the trend. We start off with potassium, K, a very active metal which reacts violently even with the relatively stable compound water. Next is calcium, Ca, which is slightly less active as a metal. Next comes scandium, Sc, and nine additional elements, all of which have very obvious metallic properties, up to and including zinc, Zn. Next comes gallium, Ga, still a metal, but with diminished metallic properties, and then germanium, Ge, still to the left of the metal-nonmetal dividing line, but now clearly ambivalent about its identity. Metals conduct electricity, but Ge does this so poorly that it is called a semiconductor, a valuable property which makes this element quite useful in constructing transistors. Arsenic, As, on the other side of the line, has definite nonmetallic properties, but can act as a metalloid. Selenium, Se, and bromine, Br, are very definitely nonmetals, as is the noble gas, krypton, Kr, which brings this horizontal row to an end. We begin the next row with rubidium, Rb, an element with properties very similar to those of K above it, and end the row with xenon, Xe, another noble gas. These obvious properties repeat themselves over and over again in moving from horizontal row to horizontal row. Therefore, the horizontal rows are called "periods" and are numbered from 1 to 7, the last period beginning with francium, Fr, and ending with element 118, Uuo, whose existence is yet to be confirmed and named.

A definite trend toward metallic character can be discerned in moving down any vertical row, called columns. For example, the first column is headed by hydrogen, H, very definitely a nonmetal. Metallic character increases with lithium, Li, sodium, Na, and so forth, until reaching francium, Fr, the most active metal known, at the bottom of the column. Similarly, moving down the column headed by carbon, C, a nonmetal, we see an increase in metallic character from silicon, Si, to germanium, Ge, a metalloid, to two definite metals, tin, Sn, and lead, Pb. The next-to-last column, headed by fluorine, F, is often called the halogen group of elements. F is the most nonmetallic of this group and hence, the most nonmetallic element. Metallic character increases in moving down the column through chlorine, Cl, bromine, Br, and iodine, I. Astatine, At., the last known halogen, is expected to be the most metallic member of this group. However, At is so rare that it has been estimated that the total amount of this element in the earth's crust comes to about one ounce. It has been so little studied that it is not yet certain if At forms At_2 molecules like the rest of the halogens.

Since each of the vertical columns contains a group of elements with similar chemical properties, these columns are sometimes called groups or families. The first two and the last six groups are called the representative elements and have been numbered with Roman numerals followed by the designation "A." Hence, the group headed by hydrogen is Group IA, beryllium's group is IIA, boron's group is IIIA, *etc.* The block of elements from scandium, Sc, to zinc, Zn, have the designation "B" following the group numeral. Scandium heads up Group IIIB, titanium IVB, vanadium VB, chromium VIB, manganese VIIB and the three groups headed by iron, Fe, cobalt, Co, and nickel, Ni are collectively designated

VIIIB. Copper, Cu, heads up IB and zinc, Zn, IIB. These elements are sometimes called the Transition Metals.

Every isotope of every element beyond bismuth, Bi (element no. 83) is radioactive. All elements beyond uranium, U (element no. 92) are known only because they have been synthesized in the laboratory. Thus, U is the last of the naturally occurring elements. Elements 57-71, called the lanthanides or rare earth elements, are so similar in chemical properties that efforts were still being made in the 1950's to separate their salts from one another. Neither technetium, Tc, nor promethium, Pm, have been found in the earth's crust, though both elements have been identified in stellar matter. Tc was the first element to be artificially synthesized, hence its name; its possibility of existence was predicted by the Periodic Table. All isotopes of both Tc and Pm are radioactive.

In summary, we can say that within each period, going from left to right, the trend in metallic character is one of a marked transition from highly active metals to less active metals to less active nonmetals to highly reactive nonmetals. This trend terminates abruptly with a noble gas. Within each group, going from top to bottom, metallic character increases. Other trends in physical properties are generally an increase in density and a decrease in melting point, again going from top to bottom, although the latter trend applies only to certain groups. The discussion of the preceding several pages is summarized diagrammatically in Figures 14.2, 14.3 and 14.4.

Although the arrangement of the Periodic Table was arrived at in the empirical fashion described above, current theories about the interior structure of the atom fully substantiate this arrangement. The Periodic Table has introduced a powerful unifying concept into the study of chemistry. It may be used in the simplest way to predict the characteristics of the

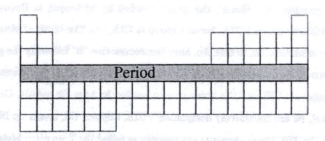

FIGURE 14.2a. The Elements in a Period Have Different Properties With Periodic Trends

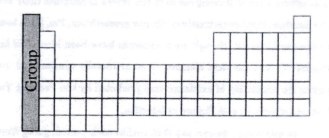

FIGURE 14.2b. The Elements in a Group Have Similar Properties With Periodic Trends

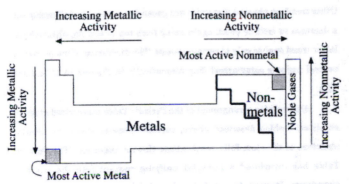

FIGURE 14.3. How Chemical Activity Varies in the Periodic Table

1. Metallic character increases from right to left and top to bottom.
2. Nonmetallic character increases from right to left and bottom to top.
3. The majority of the elements are metals.

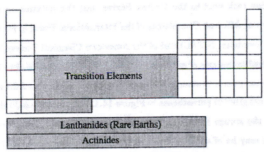

FIGURE 14.4. The Transition Elements, Rare Earths and Actinides Are Metals

elements, but it has proven to have much greater significance than this. As chemistry has evolved, the structure of the Table has correlated with each new theoretical development, thus revealing that it is based on the most fundamental theoretical concepts.

14.4 Postscript

The "A" and "B" designation given to the groups in the Periodic Table was a pedagogical device invented by a chemist, H.G. Deming, in 1923. His designation, which he incorporated into several of his textbooks, took hold in the United States, but the notation in other countries was different. Committees of the International Union of Pure and Applied Chemistry (IUPAC) and of the American Chemical Society (ACS) have worked on systems of universalization of the Table, but this has led to both non-acceptance in some circles and confusion in all. The ACS group numbers are given in parentheses in Figure 14.1. Note that the ACS simply numbers the groups from 1 to 18 right across the Table.:

14.5 Selected Readings

For a compendium of Periodic Tables from Mendeleev to the present, see http://www.meta-synthesis.com/webbook/35_pt/pt_database.php?PT_id=36 0

Femelius, W.C.; Powell, W.H. Confusion in the Periodic Table of the Elements. *Journal of Chemical Education* **1982**, *59*, 504-508.

Fluck, E. "New Notations in the Periodic Table," 1988: http://www.iupac.org/publications/pac/1988/pdf/6003x0431.pdf

Scerri, Eric. The Periodic Table: Its Story and Its Significance. Oxford University Press: New York, 2006

Chapter 15

Electron Configurations

The classification of the elements
Has not only a pedagogical importance,
As a means for more readily learning
Assorted facts that are systematically
Arranged and correlated, but it also has a
Scientific importance, since it discloses new
Analogies and hence opens up new routes
For the exploration of the elements.

Dmitri Mendeleev, 1871

CHAPTER 15

ELECTRONIC CONFIGURATIONS

15.1 Bohr Energy Levels

In the previous chapter, we learned that the elements can be placed in an order which corresponds with observed trends in physical and chemical properties, and that this order is based upon the atomic number of each element. Figure 14.1, the modern Periodic Table, is simply an arrangement which shows the regular progression of the elements in order of increasing atomic number, with a new horizontal row (period) begun for each repetition of an observed chemical property. We also observed that the atomic number, not the atomic weight, was the crucial factor in determining an element's place in the Table. Let us take a closer look at this all-important number and discover why it is such a crucial factor.

You recall from the discussion in Chapter 7 that the atomic number was defined as the number of protons in the nucleus of the atom and that it was determinative of the identity of a particular element. For example, all atoms with 40 protons in their nuclei are atoms of zirconium, Zr. Removal of a proton from a Zr nucleus (a very difficult thing to do, by the way) drops the atomic number to 39 and yields an atom of yttrium, Y, plus an atom of hydrogen (the removed proton). The reason why this is so depends upon the atomic structure of the neutral atom. You know that each proton in the nucleus carries a +1 charge, so if the atom is to be a neutral atom with a net charge of zero, it is necessary for the atom to possess an equal number of negative charges, and therefore, an equal number of electrons, each with a -1 charge. Thus, an atom of hydrogen with one proton in its nucleus will also, as the neutral atom, possess one electron outside its nucleus. The electron, if the atom is in the ground state, will occupy the lowest possible energy level shown in Figure 8 1. Likewise, an atom of helium, He, will possess two protons in its nucleus and two electrons outside its nucleus. Thus, we see that the atomic number plays a dual role: it represents the number of protons in the nucleus of the atoms of a particular element, and also the number of electrons outside the nucleus (sometimes called "extranuclear electrons") in the neutral atom. The factor that determines how a particular element will behave is the arrangement of these extranuclear electrons.

Since electron arrangement will determine elemental properties, then electron arrangement must also be very much related to how the elements are arranged in the Periodic Table. Let us see if we can correlate these two pieces of information.

We learned in Chapter 8 that electrons in atoms may occupy only certain allowed energy levels, and may not possess any energies in-between. The Danish physicist, Niels Bohr (1885-1962), first recognized this principle for the single electron of the hydrogen atom. Since energy level is associated with the probability of finding an electron in a given volume, we can conveniently develop a rough picture of the most probable volumes in which electrons may be found in terms of concentric spheres. Figures 8.5 and 8.6 depict the most probable distance of hydrogen's ground-state electron from the nucleus. If you can visualize the circle of Figure 8.6 as a three-dimensional sphere rather than a two-dimensional circle, this it is the surface of that sphere which will be the most likely place of finding the electron. However, if I were to give the electron a "hotfoot" and boost it up to the second allowed energy level, as depicted in Figure 8 1, then you can imagine that the surface of that new sphere would be much larger, and that the surface of the sphere associated with the third allowed level would be larger still. Figure 15.1 depicts these concentric spheres, but please note that this is not a picture of the atom. It is merely a representation of probabilities, and in this case, a representation of regions where the electron would most likely be found.

Although Bohr developed his theory of energy levels for the electron in a hydrogen atom, his ideas can be readily extended to multi-electron atoms. In the following discussion, we shall be using a very simplified model of the atom based on these concepts. If you go further in the study of chemistry, you will have to amend this model, but for our purposes, it is adequate.

Scientists have found that in multi-electron atoms, only a specific number of electrons can fit into any given energy level, and this number increases in going from lower levels to higher levels. If we label each principal energy level in which we can place electrons as 1, 2, 3, 4, and so forth, starting with the lowest level, and use the symbol "n" to designate these principal energy levels, then the maximum number of electrons that can fit into each level is given as $2n^2$. Sometimes "n" is called the "principal quantum level" since electrons can only absorb or give off energy in packets called quanta. Table 15.1 shows how the total possible number of electrons increases with increasing principal quantum level. The energy differences between these levels have already been diagrammed in Figure 8.1. As you might expect, the larger the spherical surface associated with a given level, the greater the maximum number of electrons that can be contained in that region.

Given an element with its full complement of electrons, an atom in the ground state will have its lowest energy levels filled first. For example, lithium, Li, with an atomic number of 3, has three electrons. Two will be found in the $n = 1$ level, which can accommodate a maximum of two electrons, while the third

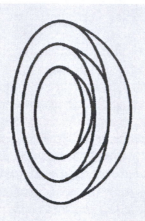

**FIGURE 15.1. Electron Level Physical
Boundary Surface
(not drawn to scale)**

TABLE 15.1
**MAXIMUM ELECTRON OCCUPANCY OF THE FIRST 6
PRINCIPAL QUANTUM LEVELS**

Principal Quantum Level, n	$2n^2$
1	2
2	8
3	18
4	32
5	50
6	72

electron must occupy the **n** = 2 state. Aluminum, Al, with 13 electrons, has 2 electrons in the **n** = 1 state, 8 electrons in the **n** = 2 state, and the remaining 3 electrons must occupy the **n** = 3 state.

Using this knowledge of the order of filling Bohr energy levels, Figure 15.2 depicts the Bohr electron configuration for the first 18 elements in the Periodic Table. Please keep in mind that this depiction is in no way meant to represent electrons circling the nucleus like planets circling the sun. This is merely a way of keeping track, or counting, the electrons in each principal quantum level.

15.2 Correlation of the Periodic Table with Electron Configuration

By examining Figure 15.2 closely, we can move quite far in our understanding of how electron arrangement correlates with the arrangement of the elements in the Periodic Table. First of all, we notice that hydrogen with one electron and helium with two are the only two elements that are contained in the first period of the Table. We also notice that only two electrons can be accommodated in the first principal quantum level. The beginning of the second period in the Table corresponds with the beginning of a new principal quantum level with lithium, and likewise, the beginning of another new period and new principal quantum level with sodium. This observed fact can be generalized for the rest of the elements in the Table: a new period marks the beginning of a new principal quantum level.

The second thing that we notice is that the number of electrons in the highest ("outermost") principal quantum level in Figure 15.2 corresponds with the group number in the Periodic Table. Hydrogen, lithium and sodium, all in Group IA of the Table, also all have only one electron in their highest principal quantum levels. Similarly, beryllium and magnesium, in Group IIA, both

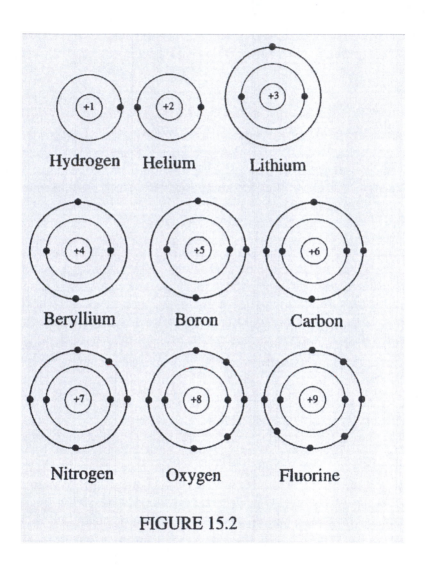

FIGURE 15.2

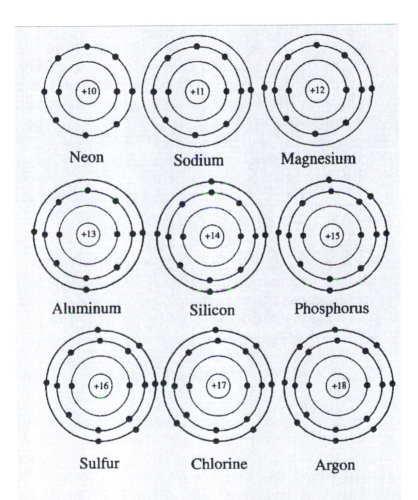

FIGURE 15.2 (continued)

have two electrons in their highest principal quantum levels, fluorine and chlorine, in Group VIIA, have seven electrons in their highest levels, and so on.

The third thing that we notice is that helium, with a complete principal quantum level, is inert. Furthermore, neon and argon, both with eight electrons in their highest principal quantum level, also belong to this category of noble gases. Therefore, having eight electrons in one's highest principal quantum level must be associated with extreme chemical stability. It was not until the mid-1950's that any of these gases could be forced to react chemically, and even then, only under extreme circumstances.

Fourthly, a decrease in metallic character is associated with an increase of electrons in the highest principal quantum level. Sodium and lithium, extremely active metals, have only one electron in that level; beryllium and magnesium, somewhat less active metals, have two; carbon and silicon, with four electrons, are no longer metals, but nonmetals, although silicon has metalloidal tendencies. The elements with 5, 6, and 7 electrons exhibit increasingly nonmetallic character.

In the fifth place, we notice that increasing metallic character is associated with movement down a group in the Periodic Table. In Group IVA, we notice that carbon is a non-metal, silicon and germanium are metalloidal, and tin and lead are very definitely metals. It seems as though metallic character depends upon the position of the electrons in the highest principal quantum level. The higher the level, that is, the farther away the electrons are from the nucleus of the atom (on average), the more likely is that atom to be a metal. We shall see later that this observation has great bearing on the chemical definition of "metal" and "nonmetal," terms we have so far defined only with respect to observed physical properties.

15.3 Electron Configurations of the Transition and Higher Elements

The electron scheme has so far taken us through the first 18 elements. Evidently a new rule is needed for the succeeding elements since potassium, with an atomic number of 19, begins a new period even though, according to Table 15.1, there is still room for ten additional electrons in the **n** = 3 level. Remember that Mendeleev knew nothing about electrons. He was arranging the sequence of known elements according to their physical and chemical properties. Our electron configuration picture must explain why this arrangement worked, and it must also lead to a scheme which readily yields the electron configuration of the element, depending upon where it appears in the Periodic Table.

You recall in our discussion of energy levels in Chapter 8 that bright line spectra could be explained on the basis of electronic transitions taking place in excited atoms. As electrons made the transition from a higher to a lower principal quantum level, the extra energy was given off in the form of electromagnetic radiation of a definite energy, and this energy could be measured. It could also be observed as a spectroscopic line if all the energies being emitted were dispersed through a slit by a prism or a grating.

Additional work by the early spectroscopists revealed the existence of additional spectroscopic lines when atoms were subjected to strong electric and magnetic fields. Since electrons are charged particles and can be disturbed by electric and magnetic fields, these new lines pointed to the existence of new electron levels, or sub-levels, within the principal quantum levels. In the presence of an electric field, spectroscopists noticed that the bright lines associated with each principal quantum level split up into new lines which they described as "sharp," "principal," "diffuse," and "fundamental." In the presence of a magnetic field, the "principal," "diffuse" and "fundamental" lines split further into 3, 5 and 7 lines respectively; the so-called "sharp" lines were the only ones that did not split further.

Decades later, when Bohr's theory was being elaborated upon, it was recognized that these splittings of spectroscopic lines had to be associated with different energy sub-levels within the atom. These levels and sub-levels have been well established, and still carry with them the original spectroscopic designations. The "sharp" lines of the spectroscopist are now called "s" sub-levels, the "principal" lines are now associated with the "p" sub-levels, the "diffuse" lines with the "d" and the "fundamental" lines with the "f" sub-levels. Further sub-levels in atoms follow the alphabet beyond f, for example, g, h, etc. The arrangement of these sub-levels for hydrogen is depicted in Figure 15.3. Note that each "s" level contains only one level; each "p" level contains 3 levels which all have the same energy unless the atom is placed in a magnetic field; all "d" levels have 5 levels, all "f" levels 7, etc. Following the rather naive picture of the atom drawn in Figure 15.2, chemists have called each principal quantum level a "shell," and so each sub-level is called a "subshell," and each of the sub-sub-levels in the subshells is called an "orbital." Thus, looking at the **n** = 2 level for hydrogen, we can say that the second principal quantum level contains two subshells, s and p. Each s subshell contains only one orbital; each p subshell contains 3 orbitals. For **n** = 3, we can say that the third principal quantum level contains three subshells, s, p and d. As with **n** = 2, the s and p subshells contain 1 and 3 orbitals respectively; the d subshell contains 5. The f subshell (not shown) contains 7.

There are 3 additional "rules" we must lay down before we can explain why potassium begins Period 4. (1) When electrons occupy energy levels, they always fill the lowest energy levels first for ground state electron configurations; (2) The maximum

205

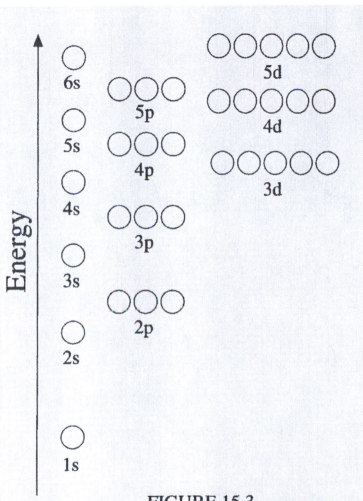

FIGURE 15.3.
Energy Level Diagram
(not drawn to scale)

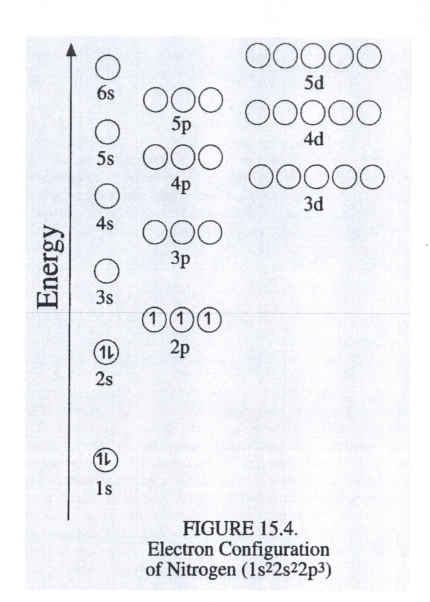

FIGURE 15.4.
Electron Configuration
of Nitrogen ($1s^2 2s^2 2p^3$)

number of electrons that can occupy any orbital is two; (3) Electrons in an atom will tend to occupy separate orbitals of equal energy when possible.

Given these rules, let us see how we can figure out the electron configuration of a simple atom such as nitrogen (At. No. = 7). Since there is only one orbital in the $n = 1$ level, this will accommodate the first two electrons. The next two electrons will enter the s-orbital of the $n = 2$ level, and the next three electrons must go into the p-orbitals of the $n = 2$ level. However, since there are three electrons to dispose of and three p-orbitals of equal energy, the electrons will occupy separate orbitals (following Rule 3, above). The arrangement will look like that of Figure 15.4. In this diagram, the electrons have been represented by arrows to indicate that they have a spin character. An arrow up and an arrow down in the same orbital circle means that the two electrons occupying that orbital have opposite spins.

From Figure 15.4, we can see that each atom of nitrogen has two electrons in the s-orbital of the first level. This can be designated in a shorthand notation as $1s^2$, where the superscript "2" indicates the number of electrons, and the number "1" indicates the shell, or principal quantum level. In the same manner, the other designations are $2s^2$ and $2p^3$. The entire configuration, then, in shorthand is $1s^2 2s^2 2p^3$. The number of electrons in the entire atom can be found by simply adding the superscripts together.

Using these same rules, we can now take potassium with 19 electrons and arrange them in the energy level diagram according to our three rules. Fill in the first, or lowest, levels first. Make sure that no more than two electrons occupy an orbital (represented as circles). If several orbitals have the same energy, such as in the p and d subshells, then several electrons entering these subshells will occupy separate orbitals. In the case of potassium, the first 18 electrons will fill the s and p orbitals of the first three shells. The 19th electron must go into the next lowest energy level which, according to our diagram, is the 4s orbital. The 4s orbital has a slightly lower energy than the 3d set of orbitals. Therefore, potassium's electron configuration is $1s^2 2s^2 2p^6 3s^2 3p^6 4s^1$. The arrangement is shown in Figure 15.5.

Similarly, element number 23, vanadium, V, will have its first 20 electrons placed to give a configuration of $1s^2 2s^2 2p^6 3s^2 3p^6 4s^2$ for the first 20 electrons; the next three electrons will enter the next lowest energy level which is now the 3d level. They will go into separate orbitals to yield the configuration shown in Figure 15.6. The complete electron configuration of vanadium then becomes $1s2 2s2 2p6 3s2 3p6 3d^3 4s^2$. Using this same method, we can see that scandium, Sc, element number 21, would have been the first element to have an electron enter a d-orbital. Since this event is unique, scandium has no precedent in the Periodic Table. Therefore, scandium's place is arranged so that there are no other elements above it in the Table. The elements from scandium to zinc (atomic numbers 21-30) are elements that are, one by one, filling in the d-subshell. As soon as we reach zinc, with 10 electrons in the d-orbitals, the next electron for element number 31, gallium, must go into the 4p subshell. Thus, gallium's electron configuration is similar to aluminum above it, and it falls right into place.

The arrangement of the Periodic Table, as may now be seen from our energy level exercises, parallels the arrangement of electrons in atoms. Groups IA and IIA are elements that are filling in s-orbitals; Groups IIIA through 0 (the noble gases) are elements that are filling in their p-orbitals. The break in the Table comes at the point where the first element to have a d-orbital occupied, scandium, comes into the picture. Thus, the entire block of elements called the "Transition Elements" are ones that are filling in their d-orbitals. Therefore, they are sometimes called the "d-

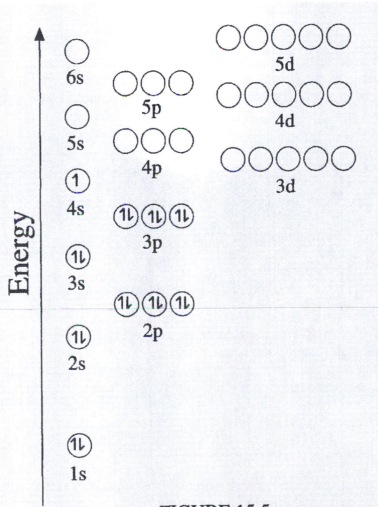

FIGURE 15.5.
Electron Configuration
of Potassium $(1s^22s^22p^63s^23p^64s^1)$

210

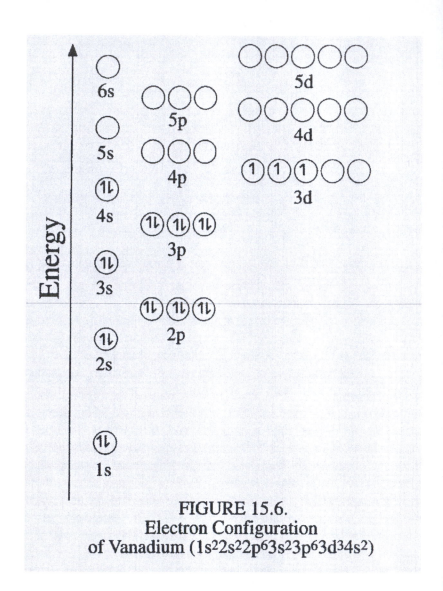

FIGURE 15.6.
Electron Configuration
of Vanadium ($1s^22s^22p^63s^23p^63d^34s^2$)

block" elements. Likewise, element number 57, lanthanum, is the first element to have an f-orbital occupied, so another break in the Periodic Table occurs at this point. However, as we pointed out previously, making room for an additional 14 elements at this point (there are seven f-orbitals) would make for a bulky and inconvenient table, so the two "f-block" groups of elements are placed separately at the bottom of the Table.

The separation of the elements in the Periodic Table into A and B groups is also based upon atomic structure. The "A" elements on either side of the Table are filling orbitals corresponding to the Period number they are in; the "B" elements are filling in the inner subshell in the next level down. (Note that "A" and "B" designations as just described have been changed by the IUPAC so that the "A" groups continue in an unbroken manner across the Table, *i.e.,* scandium heads up Group IIIA, and boron's group becomes Group IIIB. While this designation is not the most pedagogically sound, it does bring U.S. notation into line with the rest of the world.) Figure 15.7 is a schematic diagram of the Periodic Table showing the correlation of electron configuration with period and group.

In the next section, we shall see how electron configuration, determined now very easily by an element's place in the Periodic Table, places restrictions on the way atoms react to combine with one another. We will see why the formula for water is H_2O and not HO or HO_2 or H_4O. We will see why sodium and the halogens always combine to give compounds with the formula NaX. We will see why some elements are distinctly metallic while others are nonmetallic, and why some behave as metalloids. We will see why some elements, such as fluorine and potassium are extremely active, while others such as neon and argon, are relatively inert. All of these observed properties can be accounted for by electron configuration, which, in turn, can be organized into a Table which correlates configuration with these properties. Perhaps you are beginning to see why this Table is the "shorthand Bible" of chemists.

15.4 Electron Configuration as an Aid to Understanding Chemical Bonding

In the last few sections, we spent much time studying the structures of atoms. We learned that electrons in atoms are constrained to occupying only certain energy levels and that the number of electrons that can occupy these levels is limited. The first principal energy level ($n = 1$) can hold a maximum of two electrons. When $n = 2$, the maximum number is 8; when $n = 3$, the maximum number is 18, and so forth, according to the $2n^2$ rule of Table 15.1. However, an important rule of thumb is this: whenever the s and p subshells for a given value of n are completely filled, a new principal energy level and corresponding row in the Periodic Table is begun. Since an s subshell can hold a maximum of two electrons and a p subshell can hold a maximum number of six, the total number of electrons needed to fill these two subshells together is eight. This means that whenever eight electrons are attained in a given principal energy level, this is the maximum number of electrons that can occupy the level until the subshells of lower levels are filled. Achievement of eight electrons in this highest or "outermost" energy level constitutes a completed octet.

The reason why we need to know something about electron configuration is because when atoms react with one another to form stable chemical compounds, in so doing they attain an outermost octet of electrons, or an outermost doublet if it is the first principal energy level that is involved. By knowing how many outermost electrons there are, the chemist has a powerful tool for predicting the formula for any combination of elements.

As we have seen earlier, the Periodic Table is our guide in determining the number of outermost electrons. By studying the Periodic Table, the following helpful rules can be deduced:

1.   *The number of outermost electrons for the elements in the A columns of the Periodic Table always correspond to the column number.*

All of the elements in Column IA (Figure 14.1) have one outermost electron. All of the elements in column IIA have two; those in column IIIA have three, etc. For the A columns, the number of outermost electrons increases from one to seven in moving from Column IA to VIIA across each period of the Table from left to right.

2.   *The number of principal energy levels of electrons increases by one per period in going from top to bottom of any column in the Periodic Table.*

In Column IA, hydrogen (Atomic Number 1, often designated as $Z = 1$) has only one principal energy level, Li ($Z = 3$) has two, Na ($Z = 11$) has three, K ($Z = 19$) has four, *etc.* Another way of stating this rule is: The number of principal energy levels in an atom corresponds to the period number. Each horizontal row or period can be numbered from one to seven, and these numbers correspond to the number of principal energy levels occupied by each element in that period. For the A elements, the column number gives the number of outermost electrons, and the period number gives the highest principal energy level. Therefore, by combining rules 1 and 2, we can find the number of outermost electrons in the highest principal energy level for the A elements. This fact will be very useful when we study chemical bonding. Figure 15.8 summarizes these first two rules.

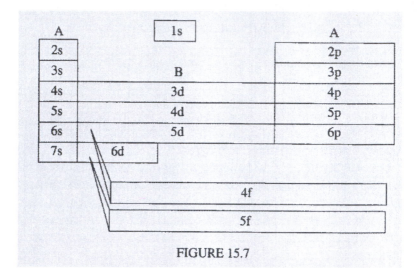

FIGURE 15.7

3. *The Transition Elements and the lanthanides and actinides all have two outer electrons with a few exceptions where there is only one electron. They differ, rather, in the number of electrons in one of the inner subshells.*

The Transition Elements (d-block elements) differ in the number of electrons in a subshell of the next inner energy level. This sub-level is called the d subshell, and the number of electrons in it is called the number of d-electrons. Scandium, yttrium and lanthanum, the elements of column IB (or IIIA, depending upon which Periodic Table you use) all have one d-electron, while the group to the right have two d-electrons, and so on up to a maximum of ten d-electrons in the column headed by zinc. The lanthanides and actinides differ in the number of electrons in the f subshell two shells below the outermost one. The f subshell can hold a maximum

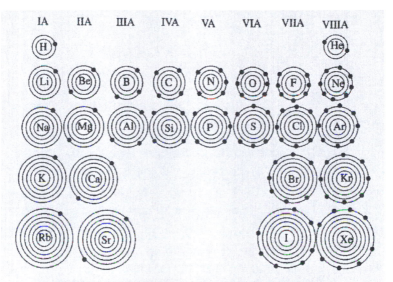

FIGURE 15.8. Periodic Variation in the Number of Outermost Electrons and in the Number of Principal Energy Levels for the Elements in the A Groups

of fourteen electrons, so there are fourteen elements in each of these series. By combining the first three rules with our knowledge about which elements are metals and which are not, we arrive at the fourth rule.

> 4. *With a few exceptions, metals have one, two or three outermost electrons. Nonmetals, with some exceptions as the number of energy levels increases, have five, six or seven outermost electrons. Helium, with only one principal energy level, has two outermost electrons. Elements with four outermost electrons are nonmetals in the lower periods but become more metallic as the number of energy levels increases.*

Now we come to the rule which is connected to chemical bonding:

> 5. *Elements with* an *outer octet of electrons are exceptionally stable and non-reactive. The helium doublet is also a uniquely stable configuration.*

15.5 Outer Octets and the Helium Doublet

A simple empirical rule about bonding is that elements form those bonds in which octets or doublets most readily result. The metals of Periods One and Two most readily form doublets. Hydrogen, in Period One, may also achieve stability by losing its single electron. To achieve these octets and doublets, atoms must somehow gain, lose or share electrons. We will discuss this further in the next chapter.

Metals tend to lose electrons. Since metals have few outermost electrons, it is energetically easier to remove these electrons than to add enough to form an octet. Metals of Columns IA and IIA which lose electrons are left with the same electron configuration as that of the nearest noble gas preceding them.

For example, lithium can lose its one outermost electron, and the remaining doublet has the same electron configuration as helium. Calcium ($Z = 20$) and barium ($Z = 56$) may each lose two electrons; the remaining configurations of their outermost electrons are those respectively of argon and xenon.

When metals lose electrons, the charged particles remaining, called positive ions, are stable. The tendency to lose electrons to form positive ions is the characteristic chemical property of metals; this property can be used as a conceptual definition of metals.

Let us see how a metal can form a positive ion using the element sodium as an example. Sodium is element number 11, which means that an atom of sodium has 11 protons in its nucleus and 11 electrons outside the nucleus arranged as shown in Figure 15.2. A glance at the Periodic Table shows us that sodium has one outermost electron which, when lost, will leave sodium with only ten electrons and the neon electron configuration. However, the loss of the electron in no way affected the number of positive charges in the nucleus, which remains at 11. Since each positive charge in the nucleus can cancel out the negative charge of each electron outside the nucleus, an atom with an equal number of each has all the charges canceled and is said to have a net charge of zero. This type of atom is called a neutral atom. However, when our atom of sodium lost an electron, a new entity with 11 positive charges and only 10 negative charges was formed. Since there is now one extra positive charge, the new entity must carry this charge. Thus, the sodium ion can be symbolized as Na^{1+} or Na^+ since the value of 1 is understood. We can depict this loss of an electron by an equation:

$$Na - e^- \rightarrow Na^+ \qquad [15.1]$$

Read: A sodium atom minus one electron yields a singly positively charged sodium ion.

Calcium, a member of Group IIA, would lose electrons thus:

$$Ca - 2e^- \rightarrow Ca^{2+} \qquad [15.2]$$

Read: A calcium atom minus two electrons yields a doubly positively charged calcium ion (with the electron configuration of the preceding noble gas, argon).

Since chemists think it is neater to have all quantities in an equation positive, we can apply our rules of algebra and move the electrons to the other side of the yield sign, which is equivalent to the "equal" sign in an ordinary algebraic equation. When we do this, however, we must change their signs and the two new equations become:

$$Na \rightarrow Na^+ + e^- \qquad [15.3]$$

$$Ca \rightarrow Ca^{2+} + 2e^- \qquad [15.4]$$

The elements of columns VA, VIA and VIIA can gain three, two and one electron respectively to achieve the electron configuration of the noble gas following them in the same period. For example, fluorine ($Z = 9$) has seven outermost electrons and must gain only one to achieve the configuration of neon. Since a gain of an electron by a neutral atom means the addition of a negative charge to an atom which originally had a net charge of zero, fluorine forms a negative ion with a single negative charge and can be symbolized by F^-. The corresponding equation is:

$$F + e^- \rightarrow F^- \qquad [15.5]$$

In like manner, sulfur ($Z = 16$) has six outermost electrons and tends to gain two more to form S^{2-}. Ions with charges of 3- are less common.

Just as we used an element's tendency to lose electrons as the conceptual definition of a metal, so we can use an element's tendency to gain electrons as the conceptual definition of a nonmetal. Summarized briefly, metals tend to lose electrons and nonmetals tend to gain them. The greater the tendency an element has to lose electrons, the more metallic it is; the greater the tendency an element has to gain electrons, the more nonmetallic it is. Metals, in losing electrons, tend to form positive ions; nonmetals, in gaining electrons, tend to form negative ions. The number of electrons lost or gained is that number sufficient to achieve an outer octet (or doublet, for Periods One and Two). Referring back to Figure 14.3, perhaps it is clearer now why fluorine is called the most active nonmetal and francium is called the most active metal. These are the two elements which have the greatest tendency to gain or lose electrons respectively.

Elements that have four outermost electrons, such as carbon (Z = 6) tend to neither gain nor lose electrons, but rather tend to share them. This property has enormous ramifications for compound formation.

In the following chapter, this discussion of electron configuration will be extended in order to learn how chemical bonds form and how the nature of the bonds can explain the properties of chemical compounds.

15.6 Selected Readings

Virtually any general chemistry textbook will enlarge upon the topics covered in this chapter with specific examples of the relationship between atomic structure and bond formation.

Langmuir, I. The Arrangement of Electrons in Atoms and Molecules. *Journal of the American Chemical Society* **1919**, *41*, 868-934. A groundbreaking paper by a Nobel laureate.

Pauling, Linus. This documentary on double-Nobel laureate Pauling summarizes his monumental contribution to our understanding of the chemical bond. http://osulibrary.oregonstate.edu/specialcollections/coll/pauling/bond/

CHAPTER 16

CHEMICAL BONDING AND COLOR

16.1 Color and Electron Transitions

In Chapters 8 and 9, we examined the electron transitions that take place in atomic and molecular systems and saw that absorption or emission of electromagnetic radiation over a certain range of energies could give rise to color. In hydrogen, for example, the energy level differences were of such magnitude that excited hydrogen atoms could emit radiation in the ultraviolet, visible and infrared regions of the spectrum, giving rise to the transitions shown in Figure 8.1. We also saw that when light from a source such as the sun, which contains the full range of visible radiation (Figure 6.2), struck a sample of unexcited, or ground state, hydrogen, radiation of the same energies emitted in the previous example could be absorbed to give rise to the dark line spectrum of Figure 9.6c. Finally, we saw that more complicated molecular systems could absorb a whole range of visible wavelengths as shown in Figure 9.6e to yield a colored object by reflection of the unabsorbed portion of the visible spectrum. It will be the purpose of this chapter to see how atoms can form more complex systems by the process of selective absorption.

16.2 The Noble Gases

A good clue to the nature of chemical bonding, which is the way atoms combine into more complex entities, comes from observation of the noble gases, helium, neon, argon, krypton, xenon and radon. Most of these gases were discovered by Sir William Ramsay

(1852-1916) and Morris Travers (1872-1961), two British chemists, around the turn of the 20th century. One of the most striking properties of these gases is their extreme lack of reactivity. Most substances, sooner or later, with or without some external help, will undergo some kind of chemical reaction. Iron rusts, gasoline burns, certain metals tarnish, fibers rot, etc. Some notable exceptions to this rule are gold, platinum, palladium and several other metals which are unreactive in the extreme. These metals were dubbed the "noble metals" because they tended to remain "aloof" from the common elements, and reacted, or "mingled" with them only under great duress. The noble gases, by extension of the term, also react only under great duress, and several of them are not known to react at all. We may conclude from this observation that there is something about the structure of the noble gases that confers great stability upon them, and stability is a state of being that all substances in the universe are striving to achieve.

An examination of the electron configuration of each of the noble gases reveals one common structural characteristic. With the exception of helium, all of them possess an outer octet of electrons. This suggests, as we pointed out in the previous chapter, that a configuration of eight outermost electrons (or two, in the case of helium) is a condition of exceptional stability. It also suggests that other substances achieve stability by achieving the electron configuration of one of the noble gases, something that can be done only by gaining, losing or sharing electrons. These processes, electron transfer or sharing, give rise to the attachments between atoms we call chemical bonds. Chemical bonding is central to the science of chemistry and also to our understanding of how various chemical substances are colored. The next three sections of this chapter will deal in turn with the concept of bonding, how some bonds are formed by sharing electrons, and how other bonds are formed by electron transfer. The final sections will then deal with some types of chemical compounds which give rise to color.

16.3 The Concept of Bonding

A bond is a tie that binds one individual or object to another in some kind of stable relationship. Family and societal bonds relate the individual person to worlds outside of himself or herself. These bonds are maintained by sensory and intellectual communication, and may sometimes be broken to form new bonds with new individuals or communities.

A chemical bond, in concept, is not very different from the idea of a societal bond. It is the means by which an individual atom relates with or becomes associated with other atoms in groups of short- or long-range stability. Formation of a chemical bond between atoms can produce several results:

1. Formation of chemical bonds produces new substances which have different properties from those of their constituent atoms.

2. When dissimilar atoms are bonded, the new substances are called compounds. When the bonded atoms are all alike, as in N_2, S_8 or P_4, the properties of the combined atoms also differ from those of the single atoms, but we have no special name for such combinations. The most stable combination of the atoms of an element at room temperature (25 $^\circ$C) and at normal atmospheric pressure is called the standard state of the element.

3. Some bonds are very unstable and hence the compounds in which they are present are very short-lived; they decompose readily and may exist for only brief periods of time. In other cases, the bonds in a compound may be quite stable and difficult to rupture; when a compound has only

such stable bonds in it, it will be long-lived under normal circumstances.

4. Depending upon the environmental conditions such as temperature and pressure, some bonds are more reactive than others upon encountering other substances. Hence, compounds, whether stable or unstable relative to their constituent elements, can always react with other substances or even rearrange themselves under the appropriate conditions. Even very stable compounds are liable to attack by very vigorous substances. Therefore, stability is subject to the environmental conditions in which a compound is placed.

When atoms unite or exchange partners to form chemical bonds, they do so by means of either sharing or transferring electrons. The number of electrons available for sharing or transfer will determine the number and kinds of bonds that an atom will form. Therefore, a knowledge of an atom's electron configuration, a subject we treated at great length in the previous chapter, is essential for understanding what will result when it interacts with other atoms to form chemical bonds. Electron configuration, when applied to bonding, will help us see why hydrogen and oxygen always combine in a ratio of two atoms of hydrogen to every one atom of oxygen to form the compound water, and why sodium and chlorine always combine in a one to one ratio to form sodium chloride, or ordinary table salt. It should also be noted that the nature and properties of the compounds formed by chemical bonding are quite different from the properties of their constituent atoms.

The concept of chemical bonding is also useful in classifying the chemical compounds that form as a result of bonding. In the early years of chemistry, attempts to classify the almost infinite variety of chemical substances resulted in the division of materials into those derived from living matter, organic compounds, and those derived from non-living matter, inorganic compounds. At first, this classification was very useful because it corresponded to observed differences in chemical and physical properties between organic and inorganic compounds. The organic compounds derived from living matter did not dissolve readily in water and they burned readily to produce gases. Inorganic materials, derived from the earth and seas, seemed impervious to fire and many of them dissolved in water. Many inorganic materials contained metals, while many organic materials contained nonmetals, particularly carbon.

As time went on, however, this classification became less satisfactory because there were so many exceptions to it. Many materials which burn readily are also soluble in water, and some organic materials do not burn. However, the event that dealt the death blow to this rather artificial classification was the accidental synthesis of urea, a product of animal metabolism, from inorganic mineral material by Friedrich Wöhler (1800-1882) in 1828. From this synthesis, it became clear that the compounds derived from living and non-living matter were interchangeable and that a different way of classifying chemical compounds was needed.

Today chemical compounds can be most conveniently classified according to the bonds they contain. When electrons are shared between atoms, the resultant type of bonding is called covalent, and when electrons are transferred, the bonding that results is called ionic. It just so happens that most compounds derived from living matter also form covalent bonds, and that many compounds derived from non-living matter form predominantly ionic bonds. Therefore, the two great divisions of chemistry still remain "organic" and "inorganic," but with modified definitions. "Organic chemistry" is usually defined now as the study of the compounds of carbon; "inorganic chemistry" is the study of the compounds of the remaining 117 elements.

16.4 Covalent Bonding: Sharing Electrons

In order for atoms to achieve a complete octet of outermost electrons, they can either share electrons or transfer them. For example fluorine ($Z = 9$) has seven outermost electrons and needs only one more to complete its octet and achieve the stable configuration of neon ($Z = 10$). When two fluorine atoms which each require one electron encounter one another, neither can easily give up an electron to the other. However, there is an arrangement that can provide both atoms with a complete electron octet. If each of the two fluorine atoms were to share one of its electrons to make up a shared pair of electrons, the shared pair would be common to both atoms and each atom would have a completed octet. This type of sharing is called covalent bonding and is depicted in Fig. 16.1a. The diagram shows that the regions in space occupied by the outermost electrons of each atom overlap, and that an electron from each atom is common to both. This common shared pair of electrons is associated with both atoms and constitutes one covalent chemical bond. The other members of fluorine's family, chlorine, bromine and iodine, also engage in this type of bonding with one another. Two atoms of chlorine are depicted sharing a pair of electrons in Figure 16.1b. The elements of Group VIIA are not the only elements that share electrons with like atoms. Hydrogen, which needs only one more electron to complete its doublet (since it is in Period One of the Periodic Table) shares electrons as depicted in Figure 16.1c. Other elements that form bonds with like atoms are oxygen, nitrogen, sulfur, carbon and phosphorus. However, this is only a partial list of elements that behave in this manner.

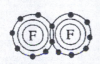 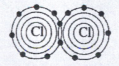

| FIGURE 16.1a. Covalent Bonding in Fluorine, F_2 | FIGURE 16.1b. Covalent Bonding in Chlorine, Cl_2 | FIGURE 16.1c. Covalent Bonding in Hydrogen, H_2 |

FIGURE 16.1. Covalent Bonding Between Like Atoms
(only valence electrons are shown)

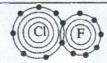

| FIGURE 16.2a. Covalent Bonding Between Chlorine and Fluorine to Form Chlorine Monofluoride, ClF | FIGURE 16.2b. Covalent Bonding Between Chlorine and Hydrogen to Form Hydrogen Chloride, HCl |

FIGURE 16.2. Covalent Bonding Between Unlike Atoms
to Form Compounds

When atoms share electrons to form a bond between them, such an arrangement is called a diatomic molecule. The sharing of electrons by two fluorine atoms to form a single entity consisting of two fluorine atoms joined by a covalent chemical bond may be symbolized by F_2. The subscript "2" indicates the presence of two atoms of fluorine; the fact that it is written as a subscript indicates that the two atoms are joined together by a chemical bond. However, since the two atoms in this grouping are of the same element, this diatomic molecule does not constitute a chemical compound. A chemical compound must contain at least two different elements. Therefore, F_2, Cl_2 and H_2 (all diagrammatically depicted in Figure16.1) are not compounds. They are elemental diatomic molecules of respectively fluorine, chlorine and hydrogen.

In order to form compounds, the sharing must take place between atoms of different elements. For example, if fluorine and chlorine were to share a pair of electrons, the arrangement would look like that depicted in Figure 16.2a. Here, one of the fluorine atoms of Figure 16.1a has been replaced by a chlorine atom. The formula for this new entity is now ClF and it is called chlorine monofluoride. This is a true chemical compound with properties different from both of the original elements which make it up. For example, ClF is a colorless gas, whereas F_2 and Cl_2 are both greenish-yellow gases. In like manner, hydrogen and chlorine can share a pair of electrons to form HCl, depicted diagrammatically in Figure 16.2b. This compound is also a colorless gas and is called hydrogen chloride. It dissolves very readily in water; a water solution of HCl is commonly known as hydrochloric acid. The properties of HCl are quite different from those of the individual elements, H_2 and Cl_2. Note also that while these examples of covalent compounds happen to be gases, many covalent compounds are liquids or solids at room temperature.

Although the diagrams in Figures 16.1 and 16.2 are helpful in visualizing a covalent bond, they are awkward to draw and contain more information than we really need to depict the bond. Since only the outermost electrons, sometimes called the valence electrons, are the ones that engage in chemical bonding, it is not necessary to depict the others. Early in the 20th century, the concept of the covalent bond was developed by a chemist named G.N. Lewis (1875-1946), and it was he who suggested that each atom could be depicted by its chemical symbol with dots arrayed around the symbol, each dot representing one valence electron. For example, an atom of fluorine could be represented thus:

In this representation, F symbolizes the fluorine nucleus as well as the inner two electrons; this is called the fluorine "kernel." The remaining seven electrons are represented by dots. Similarly, the atoms of hydrogen, chlorine, oxygen, nitrogen, and carbon can be represented respectively as follows:

$$H \cdot \quad : \overset{\cdot\cdot}{\underset{\cdot\cdot}{Cl}} \cdot \quad : \overset{\cdot}{\underset{\cdot\cdot}{O}} \cdot \cdot \quad \cdot \overset{\cdot\cdot}{\underset{\cdot}{N}} \cdot \quad \cdot \overset{\cdot}{\underset{\cdot}{C}} \cdot$$

Note that in each case, the number of dots equals the number of valence electrons and also the group number of the element. Thus, the Periodic Table is an excellent guide for drawing the Lewis dot symbols for the elements. Referring to Figure 14.1, try drawing the Lewis dot symbols for sulfur (S), sodium (Na), aluminum (Al), arsenic (As) and iodine (I).

Lewis dot notation is also handy for representing the bonding in molecules of elements and compounds. The sharing of the pair of electrons in F_2 and Cl_2 depicted in Figures 16.1a and b can be much more easily represented as shown below

$$: \overset{\cdot\cdot}{\underset{\cdot\cdot}{F}} : \overset{\cdot\cdot}{\underset{\cdot\cdot}{F}} : \quad\quad : \overset{\cdot\cdot}{\underset{\cdot\cdot}{Cl}} : \overset{\cdot\cdot}{\underset{\cdot\cdot}{Cl}} :$$

since this notation provides all the necessary information more economically. Likewise, H_2, ClF and HCl can be represented by

$$H : H \quad\quad : \overset{\cdot\cdot}{\underset{\cdot\cdot}{Cl}} : \overset{\cdot\cdot}{\underset{\cdot\cdot}{F}} : \quad\quad H : \overset{\cdot\cdot}{\underset{\cdot\cdot}{Cl}} :$$

Try drawing Lewis dot structures for Br_2 and BrF.

Once the Lewis dot structure is drawn, the completion of the outer octet (or doublet) of each atom in the structure can be easily seen. However, the electrons of greatest importance to the chemist are those which are shared by the atoms because each pair of shared electrons constitutes a chemical bond. Some chemists prefer to use the system whereby a shared pair is represented by a dash instead of two dots, and the unbonded electrons are simply omitted. In this system, H_2 can be represented as H-H and Cl_2 as Cl-Cl.

Up to this point, we have confined our discussion to diatomic molecules. Each of the examples showed how atoms could achieve completed octets by sharing a pair of electrons with one other atom. However, there are many covalent compounds which contain many more than just two atoms. For example, when oxygen and hydrogen combine to form water, the compound formed has the formula H_2O. This formula shows that in water, there are two atoms of hydrogen and one atom of oxygen per molecule. How does the octet rule bring about this arrangement?

Oxygen, in Group VIA, has six valence electrons and can

$$: \overset{\displaystyle \cdot}{O} \cdot$$

be depicted as and hydrogen, in Group IA, has only one valence electron and looks like H. Oxygen needs two more electrons to complete its octet; hydrogen needs one more electron to complete its doublet. Since each hydrogen atom can contribute only one electron to the oxygen, then two atoms of hydrogen can share in one of the oxygen's electrons. The resulting dot structure for

$$H : \overset{\displaystyle \cdot\cdot}{O} : H$$

water is The outer octet of oxygen is complete and the doublet of each hydrogen is also complete. Each hydrogen forms a single covalent chemical bond with oxygen, so water can also be depicted as H-O-H. The familiar formula for water, H_2O, is thus a result of electron sharing which brings about completion of outermost electron shells for the atoms involved. [It is important to note that these dot depictions carry no information about the actual geometric structure of the molecule, but only how the atoms are related to one another by bonding.] Since water contains three atoms per molecule, it is called a triatomic molecule. Molecules containing more than three atoms may be called polyatomic ("poly" = many) molecules.

Carbon is an example of an element that forms numerous polyatomic molecules. Carbon, in Group IVA of the Periodic Table,

$$\cdot \overset{\displaystyle \cdot}{C} \cdot$$

has four valence electrons and can be represented as $\cdot$; it can utilize these four valence electrons to form numerous compounds containing single, double and triple bonds. These bonds may be formed between other atoms of carbon or between carbon and atoms of different elements. There is such variety to the number of carbon compounds that can be thus formed that the study of the compounds of carbon alone constitutes one of the major branches of chemistry. The year 1980 saw the recording of the five millionth known compound of carbon by the *Chemical Abstracts Service* of the American Chemical Society. As of mid-2012, almost 20 million compounds have been registered, and about 15,000 are added each day. In the following discussion, we will examine several of the compounds of carbon, but will save a more complete discussion of carbon for sections 16.6 and 16.7.

Because carbon has four valence electrons, it needs an additional four electrons to complete its octet. Since hydrogen atoms can each contribute one electron to form a covalent bond, it follows that if carbon combines with hydrogen, one atom of carbon will need to bond with four atoms of hydrogen. We can picture the four atoms of hydrogen approaching the carbon thus:

By sharing four pairs of electrons so that each atom contributes one electron to each shared pair, each atom achieves stability. In the process, four covalent chemical bonds are formed to produce

$$
\begin{array}{c}
\text{H} \\[-2pt]
.. \\[-6pt]
\text{H} \vdots \text{C} \vdots \text{H} \\[-6pt]
.. \\[-2pt]
\text{H}
\end{array}
$$

which may also be depicted by dash notation as

$$
\begin{array}{c}
\text{H} \\
| \\
\text{H}-\text{C}-\text{H} \\
| \\
\text{H}
\end{array}
$$

and by a condensed formula as CH_4. This compound of carbon and hydrogen is called methane. It is the simplest of a whole host of possible compounds of carbon and hydrogen, several of which we will discuss in the following paragraphs.

Before continuing this discussion, it should be noted that dash or electron dot drawings are called structural formulas; they show how all of the atoms in a molecule are connected to each other. Structural formulas, however, do not show the relative distances between atoms, nor do they depict the actual three-dimensional shape of the molecules they represent. They merely show which atoms are linked to which, and the lengths of dashes and sizes of dots are based upon convenience only. Formulas showing only the numbers of each kind of atom in a molecule, such as CH_4, are called molecular formulas.

If the element combining with carbon were changed from hydrogen to chlorine, the compound formed when four atoms of chlorine approach a atom of carbon would be CCl_4, carbon tetrachloride, with a dot formula of

$$
\begin{array}{c}
\text{Cl} \\[-2pt]
.. \\[-6pt]
\text{Cl} \vdots \text{C} \vdots \text{Cl} \\[-6pt]
.. \\[-2pt]
\text{Cl}
\end{array}
$$

where each of the five atoms involved in the bonding has achieved a completed octet. The dash notation for this compound is

$$\begin{array}{c} \text{Cl} \\ | \\ \text{Cl} - \text{C} - \text{Cl} \\ | \\ \text{Cl} \end{array}$$

Carbon is also capable of sharing electrons with itself so that a possible compound of carbon and chlorine might be

$$\begin{array}{cc} \text{Cl} & \text{Cl} \\ | & | \\ \text{Cl} - \text{C} - \text{C} - \text{Cl} \\ | & | \\ \text{Cl} & \text{Cl} \end{array}$$

As you can see, the elimination of all dots in the dash formula makes the formula less cluttered and therefore less confusing. The advantage of the dot formula is to show that the compound containing two carbon atoms, six chlorine atoms and a total of seven covalent chemical bonds has a completed octet around every atom and that each pair of atoms is sharing two electrons. In using the dash formulas, it is important to keep in mind that one dash represents the sharing of two electrons. The molecular formula for the compound shown above is C_2Cl_6. If the chlorines were replaced by hydrogens, the new compound, called ethane, would have a molecular formula of C_2H_6 and a structural formula of

$$\begin{array}{cc} \text{H} & \text{H} \\ | & | \\ \text{H} - \text{C} - \text{C} - \text{H} \\ | & | \\ \text{H} & \text{H} \end{array}$$

This very brief look at covalent bonding with respect to carbon compounds gives a glimpse into the possibilities of much more complex arrangements. However, keep in mind that in every case, carbon can form, at most, four bonds since it has four valence electrons. Since oxygen needs two electrons to complete its octet, it will form two covalent bonds; hydrogen, in need of only one electron to complete its doublet, will form one covalent bond; the halogens, fluorine, chlorine, bromine and iodine, each in need of one electron to complete its octet, will also form one covalent bond; nitrogen will form three covalent bonds, and sometimes five. The number of bonds formed by each of the elements in question ultimately has its basis in the electron arrangements of their atoms. This arrangement can be deduced from an element's position in the Periodic Table, and the number of covalent bonds likely to form can be predicted accordingly.

16.5 Ionic Bonding: Forming Charged Particles by Gaining and Losing Electrons

The sharing of electrons is typical of nonmetals bonded to other nonmetals, and for this reason, the examples chosen for discussion in section 16.4 involved nonmetals only. In fact, whenever nonmetals bond to other nonmetals, it can be assumed that the bonding is covalent.

Metals, on the other hand, are elements which tend to lose electrons during chemical reactions. The degree to which an element tends to lose electrons is a measure of its metallic character. The most metallic elements, that is, those elements that tend to lose electrons most readily, are found at the extreme left of the Periodic Table. Let us consider the octet rule again to see why this is so.

All of the elements in Group IA have a single outermost electron, which, if lost, will allow each of these elements to achieve a complete octet. The two exceptions to this statement are lithium $(Z = 3)$ which forms a doublet, and hydrogen, which ends up with no electron at all. As we saw at the end of the previous chapter, loss of an electron causes the formation of a charged particle with a single positive charge called a positive ion. Therefore, the elements of Group IA ordinarily react to form ions with a charge of 1+: H^+, Li^+, Na^+, K^+, Rb^+, Cs^+, Fr^+.

Similarly, all of the elements of Group IIA react to lose two electrons to form ions of charge 2+ because there are now two more protons than electrons in each: Be^{2+}, Mg^{2+}, Ca^{2+}, Sr^{2+}, Ba^{2+}, Ra^{2+}. In like manner, the metals of Group IIIA tend to lose three electrons to form triply positively charged ions such as Al^{3+}. In summary, all of the metals in the A columns of the Periodic Table react by losing their "column number of electrons" to form positive ions with charges also equal to the column number.

Chemists use the term "oxidation" to describe loss of electrons. An element that loses electrons becomes oxidized, and the charge it acquires is called its "oxidation number." Thus, the elements of Group IA all achieve oxidation numbers of 1+ when they react; the elements of Group IIA achieve oxidation numbers of 2+, Group IIIA elements achieve 3+, and so forth.

By losing one, two or three electrons as the case may be, metals can achieve the electron configuration of the noble gas of the preceding period in the Periodic Table. Since such a configuration is one of great stability, metals tend to lose electrons readily and to regain them with great difficulty. In the following figure, the electron arrangements of Na^+, Mg^{2+}, Al^{3+} and Ne are shown. Each of these four species has the same number of electrons, but a different number of protons in the nucleus. Similarly, K^+, Ca^{2+} and Ga^{3+} have the same electron configuration as argon, Ar.

When atoms lose electrons to form positive ions, the electrons do not just disappear. They must be transferred to some other atomic entity. Since nonmetals need one or more electrons to

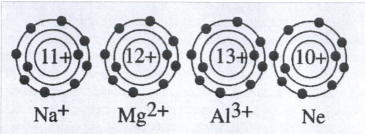

Na^+ Mg^{2+} Al^{3+} Ne

FIGURE 16.3. Electron Configurations of Na^+, Mg^{2+}, Al^{3+} and Ne.

complete their outer octets, they are likely candidates to gain the electrons lost by metals.

All of the electrons of Group VIIA need only one electron in order to achieve the noble gas configuration of the noble gas of the next higher atomic number. If chlorine, with seven electrons in its outer shell, gains one electron, it will achieve the electron configuration of argon. Similarly, fluorine will have the same electron configuration as neon when it gains one electron; bromine will have the configuration of krypton and iodine will have that of xenon. Since there is now one electron in excess of the number of protons in the nucleus of each of these examples, the ions formed all have a charge of 1-. The elements of Group VIIA are collectively called the halogens; they all react with metals to gain an electron to form ions with a charge of 1-: F^-, Cl^-, Br^-, I^-.

Each of the elements of Group VIA can complete its octet by gaining two electrons to form ions with a charge of 2-. The elements in Group VA, some of which may either gain or lose electrons, and are therefore called metalloids, gain three electrons when acting as nonmetals to form ions with a 3- charge.

To recapitulate, elements with one, two or three outer electrons can lose these electrons to form positive ions. Elements with five, six or seven outer electrons can gain electrons to form negative ions. The more likely an element is to lose electrons, the greater its metallic character; the more likely an element is to gain electrons to form negative ions, the greater its nonmetallic character. The number of electrons gained or lost is that number sufficient to arrive at an outer octet or, if no other electrons are in inner levels, a doublet. The hydrogen atom is unique in that it can lose its single outer electron to become a bare proton symbolized as H^+.

When ions of opposite charge are formed in a chemical reaction, they attract each other in the same way that electrons and protons are attracted toward each other. Together, these ions form a neutral particle held together by the strong attractive forces of the charged ions. Such linkages are called ionic or electrovalent bonds. Ionic bonds are normally formed between the metallic and nonmetallic elements. When a metallic ion bonds with a nonmetallic ion, the product is described as a salt. Thus, when K^+ combines with Cl^- to form KC1, the product is a salt called potassium chloride. Other examples of salts are common table salt, sodium chloride, NaCl, calcium iodide, CaI_2 and magnesium sulfide, MgS.

We can predict the formulas of compounds made up of the elements from the A columns of the Periodic Table. The formulas are based upon the fact that all compounds, whether they contain ions or not, must be electrically neutral, that is, the number of positive charges must be exactly balanced by the same number of negative charges. Utilizing Table 16.1, which lists the charges normally acquired by some common metals and nonmetals when they bond, we can put positive and negative ions together in the proper ratio to achieve electroneutrality. The following rules will be helpful:

TABLE 16.1. SOME COMMON IONS

A^+		B^{2+}		C^{3+}	
Ammonium	NH_4^+	Barium	Ba^{2+}	Aluminum	Al^{3+}
Copper(I)	Cu^+	Cadmium	Cd^{2+}	Arsenic(III)	As^{3+}
Hydrogen	H^+	Calcium	Ca^{2+}	Chromium(III)	Cr^{3+}
Lithium	Li^+	Chromium(II)	Cr^{2+}	Cobalt(III)	Co^{3+}
Potassium	K^+	Cobalt(II)	Co^{2+}	Iron(III)	Fe^{3+}
Silver	Ag^+	Copper(II)	Cu^{2+}		
Sodium	Na^+	Iron(II)	Fe^{2+}		
		Lead	Pb^{2+}		
		Manganese	Mn^{2+}		
		Mercury	Hg^{2+}		
		Zinc	Zn^{2+}		

D^-		E^{2-}		F^{3-}	
Acetate	$C_2H_3O_2^-$	Carbonate	CO_3^{2-}	Arsenate	AsO_4^{3-}
Bromide	Br^-	Chromate	CrO_4^{2-}	Arsenite	AsO_3^{3-}
Chloride	Cl^-	Dichromate	$Cr_2O_7^{2-}$	Nitride	N^{3-}
Cyanide	CN^-	Oxide	O^{2-}	Phosphate	PO_4^{3-}
Fluoride	F^-	Selenide	Se^{2-}		
Hydride	H^-	Sulfide	S^{2-}		
Hydrogen carbonate*	HCO_3^-	Sulfate	SO_4^{2-}		
Hydroxide	OH^-	Sulfite	SO_3^{2-}		
Iodide	I^-	Thiosulfate	$S_2O_3^{2-}$		
Nitrate	NO_3^-				
Permanganate	MnO_4^-				

General Formulas from the Table Above		
AD	BD_2	CD_3
A_2E	BE	C_2E_3
A_3F	B_3F_2	CF

1. Always write the positive ion first and the negative ion second.

2. Treat the cluster or groups of atoms in the table as if they were single atoms. The atoms in these groups are bonded together covalently and the groups themselves are sometimes called radicals.

3. Follow the formulas at the bottom of the table which show the ratios in which each of the ions in each of the columns combine with one another. These formulas arise from the balance of charge that must be maintained. For example, all of the ions in column A have a charge of 1+ and all of the ions in column D have a charge of 1-. The ions in these two columns will always combine in a ratio of one-to-one yielding the general formula AD. Similarly, all of the ions in column E, with 2- charge, will need two of the ions in column A to balance the charge; hence the formula A_2E. The remaining formulas follow the same rule.

4. Whenever a formula contains more than one of a given kind of radical, the formula for the entire radical is placed in parentheses and a subscript designating the required number is placed to its right. For example, ammonium ion, NH_4^+, is in the A column and sulfide ion, S^{2-}, is in the E column, requiring the formula of A_2E. Since two ammoniums are needed for every sulfide, the formula becomes $(NH_4)_2S$.

5. The charges on the individual ions are dropped when they are written in a complete formula because if the formula is correct, the negative and positive charges all cancel one another.

6. The compound designated by the completed formula is named by naming the positive ion first and the negative ion second. For example, Fe_2O_3 is a combination of Fe^{3+},

called iron(III) (say "iron three"), and O^{2-} called oxide. Therefore the name of this compound is iron(III) oxide (say "iron three oxide"). FeO is called iron(II) oxide (say "iron two oxide"). Why?

The following formulas were derived from Table 16.1: As_2S_3, $Ca(HCO_3)_2$, $PbCrO_4$, $Na_2S_2O_3$. Check these formulas to verify that they follow the rules for maintaining electroneutrality, and then try naming the compounds they represent. You will encounter some of these compounds later on when we discuss pigments and photographic chemistry.

16.6 Covalent Compounds of Carbon

Now that we have some idea of what ionic and covalent bonding involve, let us examine some compounds from each of these large groups to see how they give rise to color. The most important groups of compounds in the covalently bonded class are the compounds of carbon. Carbon-containing compounds produce the brilliant colors of our natural and synthetic dyes as well as the colors of many plant and animal substances such as the orange color of carrots, the red color of tomatoes and the green in green plants.

The element carbon is the principal component of all animal and plant life on our planet. Although it only comprises 0.03% of the earth's crust by weight, it is concentrated in plant and animal bodies and in the remains of once living plants and animals.

Carbon is the lightest element in Group IVA of the Periodic Table. It is the most likely of all the elements to bond covalently. Since it has four valence electrons, it can share them in four covalent bonds. Of great significance is the fact that carbon can share electrons with other carbon atoms to form long chains such as

$$: \overset{\displaystyle \cdot}{\underset{\displaystyle \cdot}{C}} : \overset{\displaystyle \cdot}{\underset{\displaystyle \cdot}{C}} : \overset{\displaystyle \cdot}{\underset{\displaystyle \cdot}{C}} : \overset{\displaystyle \cdot}{\underset{\displaystyle \cdot}{C}} : \overset{\displaystyle \cdot}{\underset{\displaystyle \cdot}{C}} : \overset{\displaystyle \cdot}{\underset{\displaystyle \cdot}{C}} : \overset{\displaystyle \cdot}{\underset{\displaystyle \cdot}{C}} : \overset{\displaystyle \cdot}{\underset{\displaystyle \cdot}{C}} : \overset{\displaystyle \cdot}{\underset{\displaystyle \cdot}{C}} :$$

A major class of carbon compounds is the class made up only of carbon and hydrogen. If the proper number of hydrogen atoms shared electrons to complete the octet of each carbon atom in the chain shown above, the compound formed would look like this:

$$H : \overset{H}{\underset{H}{C}} : \overset{H}{\underset{H}{C}} : \overset{H}{\underset{H}{C}} : \overset{H}{\underset{H}{C}} : \overset{H}{\underset{H}{C}} : \overset{H}{\underset{H}{C}} : \overset{H}{\underset{H}{C}} : \overset{H}{\underset{H}{C}} : \overset{H}{\underset{H}{C}} : H$$

Or in dash notation:

$$H-\overset{H}{\underset{H}{C}}-\overset{H}{\underset{H}{C}}-\overset{H}{\underset{H}{C}}-\overset{H}{\underset{H}{C}}-\overset{H}{\underset{H}{C}}-\overset{H}{\underset{H}{C}}-\overset{H}{\underset{H}{C}}-\overset{H}{\underset{H}{C}}-\overset{H}{\underset{H}{C}}-H$$

The four simplest hydrocarbons are methane, ethane, propane and butane, with the respective formulas:

$$H-\overset{H}{\underset{H}{C}}-H \qquad H-\overset{H}{\underset{H}{C}}-\overset{H}{\underset{H}{C}}-H \qquad H-\overset{H}{\underset{H}{C}}-\overset{H}{\underset{H}{C}}-\overset{H}{\underset{H}{C}}-H \qquad H-\overset{H}{\underset{H}{C}}-\overset{H}{\underset{H}{C}}-\overset{H}{\underset{H}{C}}-\overset{H}{\underset{H}{C}}-H$$

However, there seems to be no limit to the number of carbons that can be linked together. Compounds like polyethylene which comprise plastic objects and containers may contain hundreds of thousands of carbon atoms linked together in long chains. In addition, the carbons need not form straight chains, but may branch out. Even butane, shown above, can be rearranged to yield

$$H-\underset{\underset{\displaystyle H}{|}}{\overset{\overset{\displaystyle H}{|}}{C}}-\underset{|}{\overset{\overset{\displaystyle H}{|}}{C}}-\underset{\underset{\displaystyle H}{|}}{\overset{\overset{\displaystyle H}{|}}{C}}-H$$

$$H-\underset{\underset{\displaystyle H}{|}}{C}-H$$

which is a different compound, with different physical and chemical properties from butane. (Note: All the carbon-carbon bonds are equivalent and have the same length in the above compound; the lengths of the dashes are merely those convenient for the drawing.) This possibility of branching introduces another kind of variability since there can be several or even many different compounds with the same number of carbon and hydrogen atoms, each with slightly different chemical and physical properties. The longer the carbon chain, the greater the number of possible arrangements of atoms. There are over four billion different compounds possible for the hydrocarbon $C_{30}H_{62}$. Compounds with the same molecular formula but with different arrangements of their atoms are called isomers. Butane and its branched counterpart shown above are examples of isomers. The branched molecule is properly called methylpropane.

No other element, under the conditions present on our planet, can even approach carbon in its tremendous variety of possible compounds. Additional variation is introduced by the fact that other elements such as nitrogen, oxygen, sulfur and phosphorus may bond anywhere along the straight or branched carbon chains, or may link two or more carbon chains together. Just on this basis alone, the compounds of carbon admit of almost infinite variation.

However, even the possibilities already discussed do not tell the whole story of carbon's versatility. There is a stable hydrocarbon with the formula C_2H_4 called ethylene. Since this compound has two fewer hydrogen atoms than its two-carbon counterpart, ethane, C_2H_6, it also has two fewer electrons that the carbons can utilize to complete their octets. There is convincing evidence that the two carbon atoms in ethylene make up for this deficiency by sharing four electrons between them to form a new type of covalent bonding. This arrangement, where each of two carbon atoms has four electrons in common, is called a double bond. The formula for ethylene in dot and dash notation is

$$H \overset{\cdot\cdot}{:} C \overset{\cdot\cdot}{::} C \overset{\cdot\cdot}{:} H \qquad\qquad H\!-\!\overset{\displaystyle H}{\underset{\displaystyle |}{C}}\!=\!\overset{\displaystyle H}{\underset{\displaystyle |}{C}}\!-\!H$$

In the dot formula, in counting the number of electrons associated with the carbon atom on the left, count both the four electrons shared with the carbon atom on the right plus the four electrons shared with the hydrogen atoms. Similarly, the electrons associated with the carbon atom on the right are the four shared with the carbon atom on the left plus the four shared with the hydrogen atoms. The shared electrons count twice. Since we previously defined a covalent bond as a shared pair of electrons, it follows that when atoms share two pairs of electrons, they must be joined together by two bonds called a double bond, as shown in the dash representation on the right, where each dash represents two shared electrons. Ethylene is the simplest of the double bond-containing hydrocarbons, but many others exist which differ in the number and location of the double bonds and in the length and degree of branching of the carbon chain. Note that the two bonds in a double bond are not equivalent. One of the bonds is weaker than the other and can be more easily broken, that is, the shared electrons can be diverted into sharing themselves with other atoms. In fact, it was the rupture of the carbon-carbon double bond which first led to its detection. Double bonds are known for other atomic combinations also. Some examples are the carbon-oxygen double bond, symbolized by $-C=O$, and the sulfur-oxygen double bond, symbolized by $-S=O$.

The double bond is not the end of carbon's versatility in bonding. It is also possible for carbon atoms to share three pairs of electrons between them to form a triple bond. The simplest triply bonded hydrocarbon is acetylene, a common industrial fuel. It has the formula H:C:::C:H in dot notation and

$$H—C\equiv C—H$$

in dash notation. Again, we see that the carbon atoms and hydrogen atoms have completed octets and doublets, respectively. Both double and triple carbon-carbon bonds are vulnerable to attack and rupture readily under the proper circumstances. Triple bonds also exist in other substances such as nitrogen, N_2, with a Lewis dot structure of :N:::N: and a dash structure of

$$:N\equiv N:$$

The nitrogen-nitrogen triple bond happens to be extremely stable. Another example of a triply bonded compound is carbon monoxide, the deadly poison that occurs in automobile exhaust fumes. It has the dot formula of :C:::O: and the dash formula of

$$:C\equiv O:$$

Finally, we must note in this discussion that the formation of carbon compounds is not limited to the formation of open chains. There are many compounds in nature where carbon occurs in rings, and these rings may contain only single bonds, one double bond or many double bonds. The common ring hydrocarbon, benzene, with the molecular formula C_6H_6, is shown on the next page. Note that each carbon octet is complete and each hydrogen doublet is complete. Also note the presence of three double bonds which alternate around the ring. Alternating single and double bonds is

a circumstance known as conjugation, an important structural feature of colored organic compounds. We will examine this feature in more detail in the next section.

16.7 Conjugated Compounds of Carbon

The term molecule is used to describe each unit of a covalent compound. When one atom of carbon combines with four atoms of hydrogen to form a unit of CH_4, methane, the unit is called a methane molecule and CH_4 is called its molecular formula. Similarly, benzene has a molecular formula of C_6H_6, which means that each molecule of benzene contains six atoms of carbon and six atoms of hydrogen. The structural formula, given at the end of section 16.6 for benzene, shows us how the six carbon and hydrogen atoms are arranged.

In Chapter 15, we saw that electrons bound to atoms occupy regions in space called orbitals. Since these regions are associated only with individual atoms when the atom remains unbonded chemically, they are more specifically called atomic orbitals. When atoms combine to share an electron pair, the contributed electrons do not continue to occupy the individual atomic orbitals of the parent atoms but rather form a new region in space which they both occupy. This new region is called a molecular orbital, appropriately enough, since it arises in the formation of molecules. Every pair of atomic orbitals utilized in bonding gives rise to a pair of molecular orbitals. Hence, the number of molecular orbitals in a given compound is equal to the number of contributed atomic orbitals arising from the atoms that comprise it. The molecular orbitals have different energies for the most part, so electronic transitions between them will involve the absorption or emission of energy. When these energy differences correspond to the energies of visible light, color in a molecule is possible. Let us see what kinds of molecules fulfill these conditions.

In 1876, a chemist named Otto N. Witt (1853-1915) observed that many colored organic compounds contained certain structural groupings. Among them were -C=C-, -N=N-, -C=O, -C=S, and the benzene ring. He called these groups **"chromophores,"** a word he coined from two Greek words, *khroma,* meaning "color," and *pherein,* meaning "to bear or produce." He also noted that the mere presence of one of these groups was not sufficient to produce color, but that multiplication of or combination of these groups in molecules would often result in a colored compound. Witt also recognized that the presence of other groups such as –OH and -NH$_2$ served to strengthen and deepen the color in a molecule, so he dubbed these groups **"auxochromes."**

Although twentieth century chemistry has been able to provide a theoretical model to explain Witt's observations on the basis of molecular orbital transitions, very little else can be added to his original idea. All colored organic compounds, whether natural or synthetic, contain a certain number of chromophores and auxochromes. Chief among these is the -C=C–C=C- grouping which may occur over and over again in a chain to produce a compound that looks like this:

This compound, which contains eight -C=C- groupings, is colorless. Color does not appear until eleven such groups are present in a compound. The coloring matter in carrots, β-carotene, has the formula shown below. It contains eleven –C=C- groups, nine in the straight chain and one in each of the rings. The presence of these groups enables β-carotene to undergo molecular orbital transitions corresponding to absorption of light in the blue-violet region of the spectrum, yielding a reflected color of orange.

Whenever an organic compound contains two or more double bonds with a single bond positioned between them, the compound is said to be conjugated. When a compound contains large numbers of these alternating single and double bond structures, such as β-carotene, it is highly conjugated.

Molecular orbital theory can provide at least a partial explanation for the appearance of color with increased conjugation in the carbon chain. The bonded electrons all along the chain occupy molecular orbitals which may be divided into what are known as bonding and antibonding molecular orbitals. The bonding molecular orbitals correspond to regions in space which fall mainly between the carbon atoms in the chain, while the antibonding molecular orbitals correspond to regions in space largely at the peripheries of the carbon atoms. The bonding electrons tend to occupy the bonding molecular orbitals since they have lower energies relative to the antibonding molecular orbitals. However, absorption of the appropriate amount of energy will promote an

248

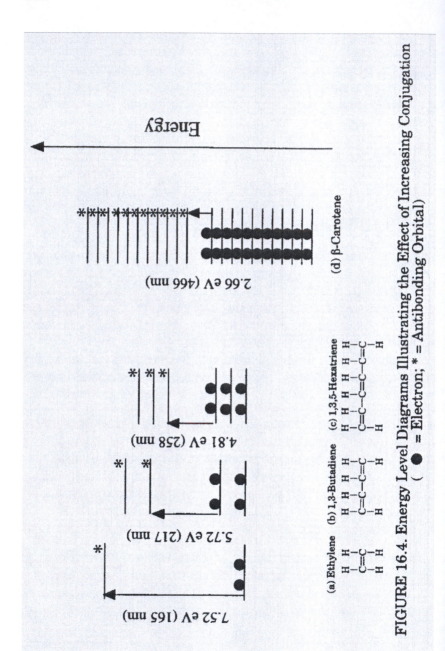

FIGURE 16.4. Energy Level Diagrams Illustrating the Effect of Increasing Conjugation
(● = Electron; * = Antibonding Orbital)

electron to an antibonding orbital. For example, Figure 16.4a illustrates the energy difference between the bonding and antibonding molecular orbitals of ethylene. (An asterisk denotes an antibonding orbital.) Ethylene has two electrons in a bonding orbital. Absorption of 7.52 electron-volts (eV) of energy, corresponding to light with a wavelength of 165 nm, will produce an excited state of ethylene by promotion of an electron to an anti-bonding orbital. Likewise, 1,3-butadiene, shown in Figure 16.4b, requires 5.72 eV (217 nm) for promotion to the excited state, and 1,3,5-hexatriene in Figure 16.4c requires 4.81 eV (258 nm) for promotion.

Several features of these figures should be noted. First, increased numbers of alternating single and double bonds, that is, increased conjugation, results in an increased number of both bonding and antibonding orbitals. Secondly, the energy difference between the highest energy bonding orbital and the lowest energy antibonding orbital decreases steadily with increasing conjugation. Thirdly, the energy differences in Figures 16.4a, b and c correspond to the energies of absorbed light in the far ultraviolet and ultraviolet regions of the spectrum. None of the compounds illustrated is colored because none of them absorbs in the visible region of the spectrum However, extended conjugation such as that in β-carotene, shown in Figure 16.4d, leads to such a small energy gap between the bonding and antibonding molecular orbitals that light with an energy of only 2.66 eV (466 nm) can be absorbed. Since this energy corresponds to that of blue-violet light, β-carotene absorbs blue-violet light and reflects orange, the complement to blue-violet. Similarly, lycopene, the red coloring matter of tomatoes and rose hips, has a structure very similar to that of β-carotene, but is modified enough to absorb in the blue-green region *(ca.* 490 nm) of the spectrum to reflect red. Other naturally occurring coloring matter with very similar structures are canthaxanthin, the red-orange colorant in edible mushrooms, which absorbs at about 482 nm, and anthacene, the red-orange colorant in cooked lobsters, which absorbs at about 500 nm. Further conjugation can lead to absorption of light of even lower energies so that, in theory, the entire energy range of the visible spectrum can be absorbed. If a compound absorbs in the red end of the visible region, for example, the color seen will be blue-green, the complement to red. The blue-green color of some sea snakes and lizards is due to red-absorbing pigmentation.

Another important chromophore is the azo group with the formula -N=N-. The many dyes that contain this grouping are called "azo" dyes, and constitute one of the largest classes of useful colorants. The presence of the azo group reduces the need for -C=C-groups to about six or seven in order to produce color. Hence, colored molecules that contain azo groups are usually smaller and less complicated than compounds that contain only the -C=C-chromophore.

Since most organic colorants consist of large, complex molecules, numerous molecular orbital transitions are possible. For example, in Figure 16.4d, eleven bonding and eleven antibonding molecular orbitals are shown. Electronic transitions may take place between any of the bonding to any of the antibonding orbitals, although the most probable transition is the one that is illustrated by the arrow. However, the fact that so many other transitions are possible means that absorption of radiation can take place over a large energy range rather than over a single narrow band or line. Thus, the colors produced by these compounds are non-ideal. Many more examples of organic colorants will be discussed in the following chapter on dyes.

16.8 Colored Inorganic Compounds

When sodium ions and chloride ions combine to form sodium chloride, the resulting compound is colorless. We may conclude that neither the sodium ion nor the chloride ion is a chromophore, that is, a color-bearing or color-producing substance. When cobalt(III) and chloride ion combine to form $CoCl_3$, cobalt(III) chloride, the resulting compound is red. The logical conclusion to draw is that the cobalt in this compound is the chromophore. Further, combination of chloride ion with each of the elements in Period Four of the Periodic Table yields the interesting results in Table 16.2. With the exception of BrCl, each of the chlorides of the Group A elements, those on the left and right sides of the Periodic Table, is either white or colorless. The chlorides of the transition (d-block) elements, on the other hand, exhibit a range of colors from the red to the violet ends of the visible spectrum. A logical conclusion derived from this observation is that there must be something special about the structure of the transition metal ions which allows them to produce color.

The electron arrangement of the transition elements that distinguishes them from the elements of the A groups in the Periodic Table is the presence of an only partially filled d-subshell. This situation is best illustrated by referring back to Figure 15.6, where we see only three electrons occupying the d-subshell of vanadium; seven additional electrons can be accommodated, thus leaving vanadium with a partially filled d-subshell. Whenever this situation occurs, molecules with pairs of electrons to contribute, such as ammonia, NH_3, and water, H_2O, can surround a transition metal ion and form bonds which involve the vacant d-orbitals of the metal ion. Such electron-pair donors are called ligands and the groups they form with metal ions are called coordinate complex ions. The number of ligands associated with a particular metal ion is called the coordination number. The coordination number has a range from two to six, ordinarily, and its value depends upon the nature of the metal ion, the oxidation state of the metal ion, and the nature of the ligand.

TABLE 16.2
COLORS OF THE CHLORIDES OF PERIOD FOUR

Formula	Name	Color
KCl	Potassium chloride	Colorless
$CaCl_2$	Calcium chloride	Colorless
$ScCl_3$	Scandium chloride	Colorless
$TiCl_2$	Titanium(II) chloride	Light brown; black
$TiCl_3$	Titanium(III) chloride	Dark violet
$TiCl_4$	Titanium(IV) chloride	Light yellow
VCl_2	Vanadium(II) chloride	Green
VCl_3	Vanadium(III) chloride	Pink
VCl_4	Vanadium(IV) chloride	Red-brown
$CrCl_3$	Chromium(III) chloride	Violet
$MnCl_2$	Manganese(II) chloride	Pink
$FeCl_2$	Iron(II) chloride	Green-yellow
$FeCl_3$	Iron(III) chloride	Orange-yellow
$CoCl_2$	Cobalt(II) chloride	Blue
$CoCl_3$	Cobalt(III) chloride	Red
$NiCl_2$	Nickel(II) chloride	Yellow
$CuCl_2$	Copper(II) chloride	Brown-yellow
$ZnCl_2$	Zinc chloride	White
$GaCl_3$	Gallium chloride	White
$GeCl_4$	Germanium tetrachloride	Colorless
$AsCl_3$	Arsenic(III) chloride	Colorless
$SeCl_4$	Selenium tetrachloride	White
BrCl	Bromine monochloride	Red to colorless

Whenever a ligand interacts with a transition metal ion, the possibility of electronic transitions within the d-subshell arises. Such transitions are called d-d transitions, and most of them require energies corresponding to the energies of visible light. For this reason, transition metal ions can exhibit color, the hue and intensity of which depends upon the nature of the ligand. The only structural requirement is that the metal ion have a partially filled d-subshell.

The transition metals most likely to exhibit color due to d-d transitions are titanium (Ti), vanadium (V), chromium (Cr), manganese (Mn), iron (Fe), cobalt (Co), nickel (Ni) and copper (Cu). Compounds of these elements exhibit brilliant colors which cover the entire spectral range. Most chromium compounds are green and blue. Manganese usually has pink to rose-colored compounds. The compounds of iron due to d-d electronic transitions range from pale green to deep red. Cobalt compounds exhibit several colors including violet, green, blue, red, pink, gray and orange. The color green is usually associated with nickel(II) compounds and the color blue with copper(II) compounds.

The remaining transition elements do not usually exhibit color due to d-d transitions. Scandium (Sc), yttrium (Y) and lanthanum (La) have no electrons occupying their d-subshell in their ordinary oxidation state of 3+, and thus do not fulfill the "partially filled" requirement. Zinc (Zn), cadmium (Cd) and mercury (Hg) have a completed d-subshell, a condition that also forbids d-d transitions. Gold (Au) and silver (Ag) both exhibit oxidation states of 1+, and in this state, they too have their full complement of d-electrons. The other transition elements are rare and many of them are precious, so they would not be likely to find their way into the list of common pigments even if their compounds did exhibit color. However, because of the increased number of electron shells in these elements (namely, zirconium, niobium, molybdenum, technetium, ruthenium, rhodium, palladium, hafnium, tantalum, tungsten, rhenium, osmium, iridium and platinum), the energies required for d-d transitions in their ions are from 25 to 50% greater than in the titanium-copper series and they absorb mostly in the ultraviolet region of the spectrum.

Electronic transition within the d-subshell is not the only way that inorganic compounds can produce color. Many of them exhibit color by absorbing the energy necessary to transfer an electron from either the metal ion to the ligand, or from the ligand to the metal ion. This process is called "charge transfer" and is responsible for many of the colors observed in the transition metal complexes as well as in compounds that cannot undergo d-d transitions. Charge transfer transitions account for the colors of the compounds of lead (Pb), arsenic (As) and antimony (Sb), to mention just a few. A special case of charge transfer involves an electronic transition from a metal ion in one oxidation state to an ion of the same metal in a different oxidation state. The iron in Prussian blue exists in two different oxidation states and can undergo the transition:

$$Fe^{3+} + e \rightarrow Fe^{2+}$$

This charge transfer transition is responsible for the intense blue color of Prussian blue.

A combination of d-d transitions and charge transfer transitions accounts for the colors of many of the common gems and minerals.

Finally, several compounds important to artists, namely vermilion (HgS, mercury(II) sulfide), CdS (cadmium sulfide) and CdSe (cadmium selenide) exhibit color due to an electron energy transition process known as "band transitions." While it is not necessary to examine the way in which compounds undergo band transitions, it is interesting to note that this mechanism gives rise to unexpected colors and that the color range is from white to yellow to orange to red to black. This mechanism never allows for violet, blue or green colors.

In Chapter 18, we will look at the compounds that constitute the common artists' pigments in greater detail.

16.9 A Comment on Notation

Many of the compounds in this chapter bear Roman numeral designations, such as iron(II) chloride and iron(III) oxide. This is necessary to distinguish between ions of the same element that can exhibit different oxidation states. The phenomenon occurs mainly in the d-block where some elements can exhibit as many as five oxidation states, such as manganese. This is due to fact that these elements can lose not only their valence electrons but varying numbers of d-electrons as well.

16.10 Selected Readings

Orna, M.V., "The Chemical Origins of Color," *Journal of Chemical Education* **55,** 478-84 (1978).

Orna, M.V., "The Chemistry Behind the Artist's Palette," *Color Research and Application* **3,** 189-196 (1978).

Plus the references given in the previous chapter.

Miniexperiment XIV

To observe an example of a colored compound that undergoes charge transfer, mix a dilute solution of Fe^{3+} as $Fe(NO_3)_3$, iron(III) nitrate, with a dilute solution of CNS^- as KCNS, potassium thiocyanate. The complex ion $FeCNS^{2+}$ is formed. Charge transfer occurs between the Fe^{3+} and the CNS^- causing the complex to absorb strongly in the 500 nm region. The complex therefore exhibits a deep red color.

Miniexperiment XV

An electric current may flow only when an electrical circuit is complete. Pure water is non-conducting, so electricity cannot flow through it (unless you can supply energy equivalent to a lightning bolt). However, when ionic compounds are dissolved in water, the presence of the ions allows the solution to conduct an electric current and to complete an otherwise incomplete circuit. This fact permits a simple test for the presence of ions when compounds are dissolved in water, and therefore, for the presence of ionic bonding.

If you do not have a conductivity test apparatus available, you can make your own by stripping a used battery of its two carbon rods. Wind copper wire around the tops of the rods and connect one rod to the outside (-) terminal of a 6-volt battery (or three 2-volt batteries connected in series). Connect the other rod to a 2.5-volt light bulb, and in turn, connect the bulb to the inside (+) terminal of the 6-volt battery. Hang the rods from a piece of cardboard so that they are positioned vertically about 2-cm apart. Place the liquid to be tested in a baby food jar and dip the rods partway into the liquid. If the bulb lights, the liquid contains ions that have permitted completion of the electric circuit. Test tap water, deionized water, alcohol, ammonia, vinegar and solutions of sugar, table salt and baking soda. As a substitute, try using the two wire leads on an LED.

Miniexperiment XVI

Prepare solutions of KCNS, $K_3Fe(CN)_6$, potassium ferricyanide, and $K_4Fe(CN)_6$, potassium ferrocyanide, and swab each of the solutions onto each of three sheets of paper using small sponges. Allow the papers to dry. Prepare a fourth solution of $FeCl_3$, iron(III) chloride, and use it as a writing ink on each of the treated sheets of paper. What do you observe?

Miniexperiment XVII

Try this one on your friends. Place a small amount of potassium ferrocyanide, $K_4Fe(CN)_6$, and a small amount of iron(III) ammonium sulfate, $FeNH_4(SO_4)_2$ on a piece of white paper. Mix the two powders well and then rub the mixture over the entire surface of the paper. Shake off the excess powder. Dip a pen point or other sharp object into water and begin writing on the paper. What do you observe?

Part IV

Practical Applications

Dyes
Pigments
Paints
Photography
Ceramics
Glasses and Glazes
Art Hazards

CHAPTER 17
DYES

17.1 General Introduction

The general term "colorant" is most correctly used to refer to a substance which may be used to impart color to an otherwise colorless object. The two main classes of colorants are dyes and pigments, two terms that are often used interchangeably, and not without some confusion. In the past, it was possible to distinguish a pigment from a dye on the following grounds: pigments were inorganic, insoluble substances which required a binder to hold the material to the surface; dyes were organic, soluble materials and did not require a binder. However, with the advances of modern color technology, these distinctions are no longer clear-cut. There are many organic pigments now on the market; some colorants are classified industrially as dyes or pigments on the basis of their use; some dyes are normally insoluble in water and are only made soluble by a chemical reaction in the dyeing process itself. It therefore seems reasonable to distinguish between dyes and pigments on the basis of need for a binder: if a colorant requires a binder in order to adhere to a substrate, it is a pigment; if it does not, it is a dye. Utilizing this distinction, many colorants will be defined as one or the other in different usage circumstances. This chapter will concentrate on those colorants ordinarily used as dyes, and the substrates, or fibers, to which they adhere.

A dye is a colorant which permeates a substrate such as a textile fiber and is anchored there by chemical attraction or physical entrapment.

All dyes are organic in chemical composition. While there are virtually no inorganic dyes, many inorganic compounds such as metal salts are used extensively in some dyeing processes.

Dyes have been known since ancient times and all dyes came from natural sources until William Henry Perkin's (1838-1907) synthesis of mauveine in 1856. This event opened the door to the manufacture of the vast array of synthetic dyes now on the market. In these days when so many synthetic colorants are easily purchased, it is hard for us to comprehend the importance of Perkin's discovery. Until he made his discovery and developed it, only a few natural vivid dyes were available and they were rare and expensive.

Some of the dyes known and used from antiquity have some interesting stories associated with them. The roots of the madder plant provided a red dye with which Egyptian mummy wrappings were treated. Alexander the Great (356-323 BCE) is said to have splashed his soldiers with this red dye to fool the Persians into thinking that his army was badly wounded. By a laborious extractive process, madder root can be made to yield a bright red dye called "Turkey Red."

Indigo, from the leaves of *Indigofera tinctoria,* was grown in India over 4,000 years ago and was a major agricultural export by the nineteenth century. The Romans who invaded England encountered the Picts who used indigo to tattoo and paint themselves. The word "Briton" is a Latin word apparently derived from a Celtic word meaning "painted men "

The story of Tyrian purple (dibromoindigo) is especially interesting. This dye, brilliant purple in color, is prepared from the secretions of the shellfish *Murex,* a species that is abundant in the eastern Mediterranean. Tyrian coins depict the legend that the dye was discovered by Hercules when his dog bit into a snail and the dog's jaws became stained with a vivid purple. Since it was estimated that over 8,000 snails were required to produce only one gram of the dyestuff, the dye was very expensive. Roman rulers reserved it for their own use, forbidding anyone outside the court to wear purple robes on pain of death. This practice gave rise to the terms "royal purple" and "born to the purple."

An important red dye, cochineal, was made from the dried, powdered bodies of the South American insect, *Dactylopius cacti.* About 70,000 insects were needed to supply 35 grams of the dye. Attesting to the importance of dyes were the facts that the ancient Peruvians measured a person's wealth in terms of dyed textiles, and that the Spanish Conquistador Hernán Cortéz (1485-1547) extorted cochineal as well as gold from the unfortunate Aztecs. There is good reason to believe that the red of the British "redcoats" was cochineal.

All of this changed when William Henry Perkin, at the age of seventeen and with only a few years of training in chemistry, discovered mauveine. During Easter week of 1856, Perkin was attempting to synthesize quinine by reacting impure aniline oil derived from coal tar with potassium dichromate. To his disgust, he obtained a black sticky mass which he discarded. Later, when he used alcohol to try to clean the container, he found that it produced a beautiful bright purple solution. He repeated the experiment using a purer aniline, but obtained no color. Eventually, he discovered that it was an impurity in the aniline, toluidine, that had led to the color. Perkin perfected the process, went into the dye manufacturing business, and became rich and famous. Although his discovery was serendipitous, Perkin deserved his fame because of the genius with which he perceived and worked out his discovery. Figure 17.1 gives the structural formula for mauveine. This is a shorthand drawing where each of the corners in the rings represents a carbon atom to which is

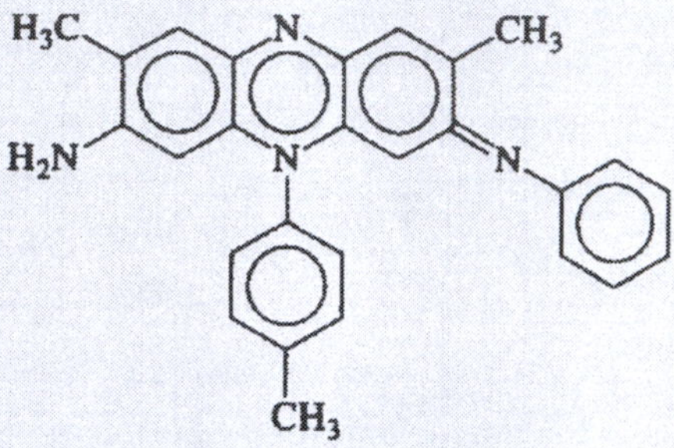

262

FIGURE 17.1. Mauveine

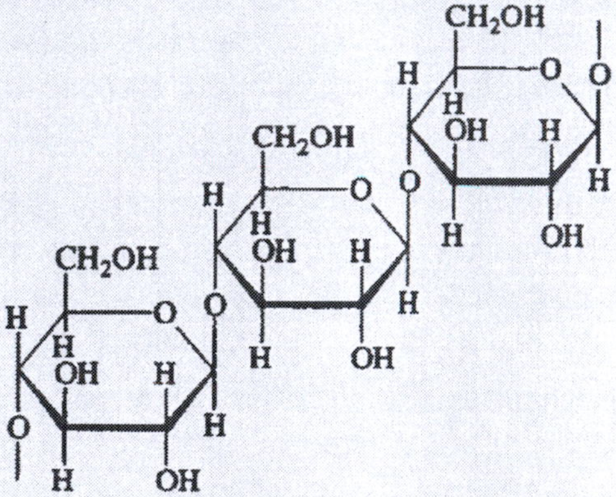

FIGURE 17.3. Part of a Cellulose Chain
(Two Rings Constitute a Cellobiose Unit)

bound the proper number of hydrogen atoms. In this formula, all of the "corners" which do not have a bond leading to another part of the molecule have one hydrogen each bonded to them.

Today, there are well over 100,000 synthetic dyes of which more than 2,000 are industrially produced. The synthetic dyes available are of such superior quality to the natural dyes and are so reproducible that commercial marketing of natural dyes has fallen to practically zero. However, many artists, weavers and hobbyists are still very interested in natural dyeing and many excellent handbooks and articles have been written on this subject. Furthermore, dye synthesis is inimical to the environment unless proper precautions are taken, and so many people prefer the use of natural dyes for this reason.

Some natural dyes and their sources are listed in Table 17.1. The natural coloring materials shown here have all been synthesized in the laboratory and are marketed today under their chemical names. Red madder, indigo and cochineal are other examples of natural colorants which are now synthesized and sold as chemicals. For example, cochineal's red coloring matter is carminic acid, a compound that can be purchased from some chemical supply houses (so "natural" is not always "natural!").

It would take many pages to list here just a small fraction of the several thousand synthetic dyes presently on the market. Even the order of listing would be quite arbitrary because dyes may be classified according to structure, that is, chromophore groupings in a molecule, or according to methods by which they are applied to the substrate, that is, usage or application. The five-volume *Colour Index,* 3rd edition, published by the Society of Dyers and Colourists in 1971, classifies 8,000 dyes and pigments under more than 40,000 commercial names. The colorants are catalogued according to structure and separately according to application class. Hue, dyeing behavior, printing behavior, lightfastness,

TABLE 17.1
SOME NATURAL DYES

Dye	Coloring Material	Color of Solid	Source
Quercetin	Quercetin	Yellow	Inner Bark of *Quercus tinctoria*
Cochineal	Carminic Acid	Red	Ground Bodies of Female *Coccus cacti*
Logwood	Hematein	Gray	Chips of Campeachy Wood (*Haematoxylon tinctorum*)

(resistance to fading upon exposure to light), and other properties are given.

17.2 Dye Substrates: Fibers

The materials to which dyes are applied are almost as numerous as the dyes themselves. In addition to textiles, where dyeing finds its heaviest use, paper, plastics, leathers and furs are all susceptible of being dyed and therefore fall into the category of dye substrates. However, we shall confine our discussion to textiles because the general principles of dye applications apply to other types of substrate as well.

The following classification of textile fibers is condensed from William Postman's *Dyeing - Theory and Practice,* a booklet published by the General Aniline Co. This list includes almost every fiber type except glass, ceramic and Teflon® fibers.

I. Natural Fibers
  A. Cellulosic - cotton, ramie, flax, jute, *etc.*
  B. Protein - wool, silk, hair
  C. Mineral – asbestos
II. Fibers Made From Natural Polymers
  A. From Cellulose - Viscose, Bemberg, Fortisan
  B. From Modified Cellulosic Fibers - Acetate, Arnel
  C. From Proteins - Vicara (from the protein zein, in corn)
III. Synthetic Fibers
  A. Polyamides - The Nylons
  B. Polyesters - Dacron
  C. Vinyl Polymers - Polyethylene, Orlon, Acrilan, Verel, Creslan, Zefran, Dynel, Vinyon, Saran, Darlan

All textile fibers are made up of long-chain polymers. Polymers are giant molecules entirely made up of "building block" units repeated over and over again in the structure. Polymers are made by combining smaller molecules, called monomers, into very long chains. For example, polyethylene, used in the manufacture of items like plastic garbage bags, is made up of long strands:

$$-\overset{\displaystyle H}{\underset{\displaystyle H}{C}}-\overset{\displaystyle H}{\underset{\displaystyle H}{C}}-\overset{\displaystyle H}{\underset{\displaystyle H}{C}}-\overset{\displaystyle H}{\underset{\displaystyle H}{C}}-\overset{\displaystyle H}{\underset{\displaystyle H}{C}}-\overset{\displaystyle H}{\underset{\displaystyle H}{C}}-\overset{\displaystyle H}{\underset{\displaystyle H}{C}}-\overset{\displaystyle H}{\underset{\displaystyle H}{C}}-\overset{\displaystyle H}{\underset{\displaystyle H}{C}}-\overset{\displaystyle H}{\underset{\displaystyle H}{C}}-\overset{\displaystyle H}{\underset{\displaystyle H}{C}}-\overset{\displaystyle H}{\underset{\displaystyle H}{C}}-\overset{\displaystyle H}{\underset{\displaystyle H}{C}}-\overset{\displaystyle H}{\underset{\displaystyle H}{C}}-\overset{\displaystyle H}{\underset{\displaystyle H}{C}}-\overset{\displaystyle H}{\underset{\displaystyle H}{C}}-\overset{\displaystyle H}{\underset{\displaystyle H}{C}}-\overset{\displaystyle H}{\underset{\displaystyle H}{C}}-$$

Each strand in polyethylene may contain thousands of CH_2 groups linked together. The monomer used to manufacture this polymer is ethylene, $H_2C=CH_2$ Each ethylene attaches itself to another ethylene by opening up one of the bonds in the double bond holding the two carbons together:

$$2\ \overset{\displaystyle H}{\underset{\displaystyle H}{C}}=\overset{\displaystyle H}{\underset{\displaystyle H}{C} \longrightarrow -\overset{\displaystyle H}{\underset{\displaystyle H}{C}}-\overset{\displaystyle H}{\underset{\displaystyle H}{C}}-\ +\ \overset{\displaystyle H}{\underset{\displaystyle H}{C}}=\overset{\displaystyle H}{\underset{\displaystyle H}{C}} \longrightarrow -\overset{\displaystyle H}{\underset{\displaystyle H}{C}}-\overset{\displaystyle H}{\underset{\displaystyle H}{C}}-\overset{\displaystyle H}{\underset{\displaystyle H}{C}}-\overset{\displaystyle H}{\underset{\displaystyle H}{C}}-}$$

This process continues until many thousands of ethylene molecules are linked together to form what is literally "many ethylenes," namely, "polyethylene." This process is known as addition polymerization. All of the synthetic vinyl polymers, such as the materials out of which auto seat covers and vinyl floors are made, are manufactured by this process.

In other cases, the monomers consist of one or two different monomers which are reactive at both ends. These monomers combine at each end to form long chains. If A and B symbolize monomers, they may combine to form a chain ....A-B-A-B-A-B-A-B... . This type of polymerization is called condensation polymerization. It also always produces another product besides the polymer for each new bond, such as water. All polyesters and polyamides are produced by this process. The formation of nylon, a synthetic polyamide, is illustrated in Figure 17.2.

FIGURE 17.2. Preparation of Nylon 6.6

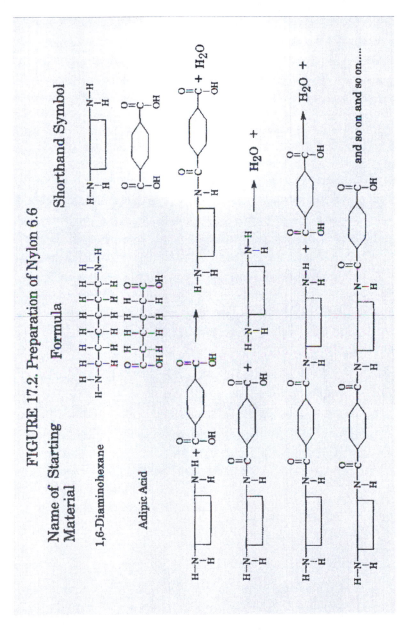

Many naturally occurring materials such as cellulose, protein, DNA, wool, silk and cotton are polymers. Cellulose fibers are composed chiefly of glucose monomers, but they are put together in a special way known as a β-1,4-linkage. In Figure 17.3 (page 262), each ring is a glucose molecule. The oxygen linking the rings together is attached to the first carbon of the left hand ring and the fourth carbon of the right hand ring, in the conventional numbering system, to give the overall molecular geometry known as "beta." The rings are drawn as if they were lying on a plane perpendicular to the plane of the paper, the dotted bonds projecting behind the plane of the paper and the bold-face bonds projecting out of the plane of the paper towards the viewer. The solidly drawn bonds are all parallel to the plane of the paper. When numerous glucose units are linked together in this way to form a long-chain polymer, the aggregate is called cellulose. Cellulosic fibers are made up of bundles of these cellulose molecules. The cellulose bundles are thought to be held together by a special form of attractive force called "hydrogen bonding." When a hydrogen on one chain is attracted to an oxygen on a parallel chain like -O-H....O- the attractive force, symbolized by the dotted line, is called a hydrogen bond. A hydrogen bond is not a true chemical bond. It is only half as strong, on the average, as a true chemical bond, but nevertheless, it is a force to be reckoned with.

When cellulosic fibers are placed in water, they swell up and thus enlarge the spaces between the cellulose chains. This allows dye molecules to enter the interstices between the chains and lodge in the fibers. Furthermore, cellulosic fibers are also hydrophilic, that is, they are water-loving, so that if water attracted to the fiber happens also to contain some dissolved dye molecules, the entry of the dye into the fiber is facilitated by this property. The amount of space between the cellulose chains per kilogram of fiber is called the internal volume. The main differences in the dyeability of the various types of cellulosic fibers is attributed to the differences in internal volume, that is, the volume available to the dyestuffs.

Protein fibers consist of bundles of polypeptide chains, polymers that have amino acids as their monomers. Amino acids have the general formula

$$R-\underset{\underset{NH_2}{|}}{\overset{\overset{H}{|}}{C}}-\overset{\overset{O}{\|}}{C}-OH$$

where R can be a number of chemical groupings ranging from H-, the simplest, to something as complex as a five-carbon chain or a ring of carbon atoms. The R- groups may also contain other groups such as acid groups, basic groups, sulfur atoms, *etc.* Given the fact that animal proteins can contain as many as eighteen different R-groupings, and that all different combinations of amino acid sequences are possible in a protein chain, protein molecules are very complicated indeed. A segment of a protein has the general formula where

$$-\overset{\overset{O}{\|}}{C}-\overset{\overset{H}{|}}{N}-\underset{\underset{R}{|}}{\overset{\overset{H}{|}}{C}}-\overset{\overset{O}{\|}}{C}-\overset{\overset{H}{|}}{N}-\underset{\underset{R'}{|}}{\overset{\overset{H}{|}}{C}}-\overset{\overset{O}{\|}}{C}-\overset{\overset{H}{|}}{N}-\underset{\underset{R''}{|}}{\overset{\overset{H}{|}}{C}}-\overset{\overset{O}{\|}}{C}-\overset{\overset{H}{|}}{N}-\underset{\underset{R^*}{|}}{\overset{\overset{H}{|}}{C}}-$$

-R, -R', -R" and –R* designate four different -R groups. The

$$-\overset{\overset{O}{\|}}{C}-\overset{\overset{H}{|}}{N}-$$

linkage is called the peptide linkage; hence the term "polypeptide" arises since many of these linkages occur in a protein chain.

Wool fiber is different from cellulosic fiber in that the forces holding the chains together consist of sulfur-sulfur linkages and salt linkages (produced by attractions between positive and negative charges on adjacent chains) as well as hydrogen bonds. Wool has been found to both absorb and liberate hydrogen ions (H^+) when placed in aqueous solution, depending upon the hydrogen ion concentration of the surrounding medium. Thus the charge on the wool fibers can be regulated by regulating the acidity of the solution, and this will affect the affinity of certain dyes for the wool. The core of the wool fiber is hydrophilic, but the outer sheath is hydrophobic.

Silk fiber is similar to wool except for the absence of sulfur-sulfur linkages, a fact that accounts for some of the physical and chemical differences between these two animal fibers.

Perhaps the most familiar fiber made from natural polymers is cellulose acetate, which goes by the name of "acetate" in everyday parlance. In 1865, the French chemist Paul Schützenberger (1829-1897) discovered that cellulose acetate could be made by reacting ordinary cellulose with a form of acetic acid. The -OH groups on the cellulose chain (see Figure 17.3) react with the acetate group to produce cellulose acetate. The chief difference between cellulose acetate and cellulose is that the former is hydrophobic, that is, it shuns water molecules; it also has a highly negative charge on the surface of the polymer chains, thus repelling any negatively charged dyestuffs. These two factors render the dyes that are normally used to dye ordinary cellulose quite unsuitable for use with cellulose acetate.

The structural formulas of several synthetic fibers are shown on the next page. Nylon 6.6 is obtained, as illustrated previously, by the condensation of adipic acid and hexamethylene diamine; the 6.6 indicates that each of the intermediates used to form the chain contains six carbon atoms. The manufacturing process of nylon yarn causes virtually all of the amide groups, which are identical to the peptide linkage shown on the previous page,

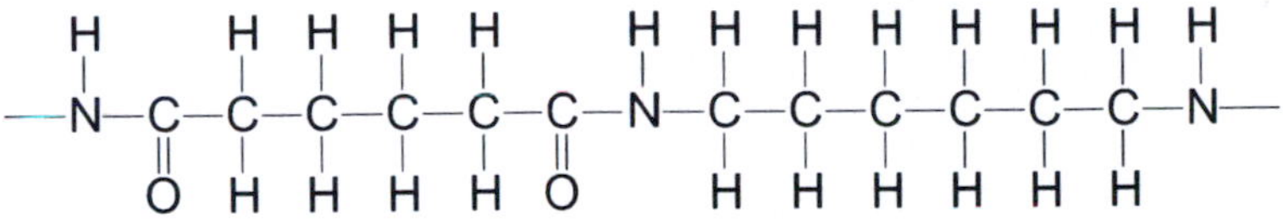

Nylon: A Polyamide

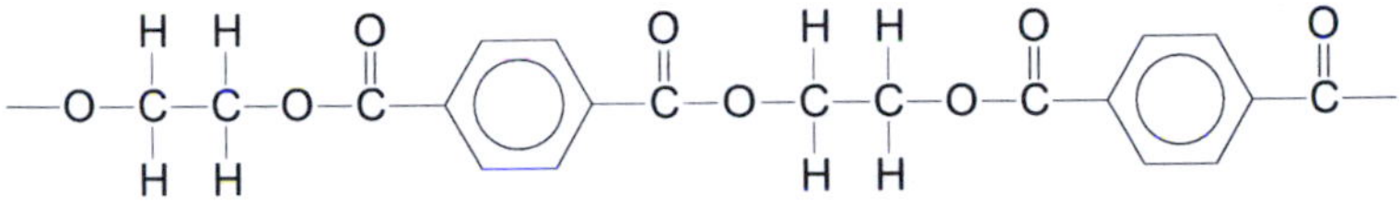

Dacron: A Polyester

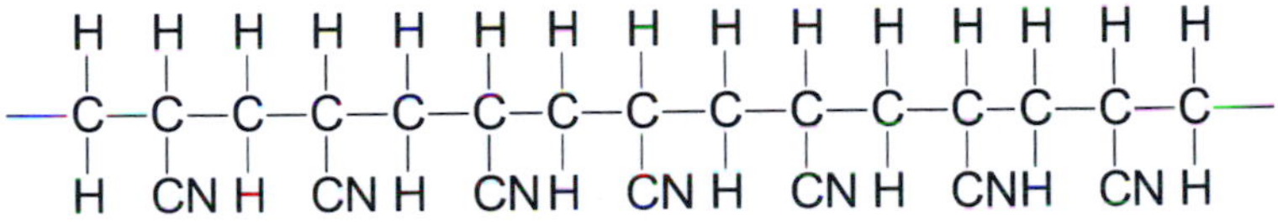

Orlon: A polyacrylonitrile

to become involved in hydrogen bonding with similar groups on neighboring chains, thus producing a hydrophobic and very difficultly dyeable material. Dacron, marketed in Great Britain as "Terylene," is a condensation product of terephthalic acid and ethylene glycol. Orlon is obtained by the addition polymerization of acrylonitrile,

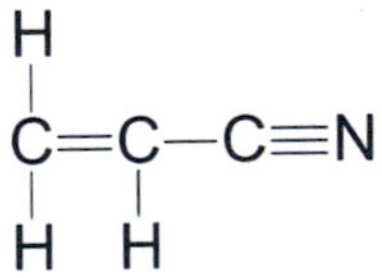

The various types of synthetic fibers are numerous and future developments in this area are limited only by the chemist's imagination. It remains for us now to see how the combined natures of dye and substrate lead to the adherence of the dye to the textile fiber. In order to do this, we must first look at the nature of the forces that bind these molecules together.

17.3 Molecular Binding Forces

There are four major types of forces which contribute to the binding of a dye to a fiber. Which of these will predominate depends upon the chemical constitutions of both dye and fiber.

A. Ionic Forces or Electrostatic Attraction (also called Salt Linkages). An ion is a charged particle which attracts ions of opposite charge and repels ions of like charge. It follows that a negatively charged substrate (fiber) will attract a positively charged dye, and vice versa. The following series of equations will show how electrostatic attraction serves to bind a negatively charged dye to wool fiber:

$$W\text{-}NH_2 + W'\text{-}COOH \rightarrow W\text{-}NH_3^+ + ...^-OOC\text{-}W' \qquad [1]$$

$$W\text{-}NH_3^+ ...^-OOC\text{-}W' + HX \rightarrow W\text{-}NH_3^+ .. X^- + W'\text{-}COOH \qquad [2]$$

$$W\text{-}NH_3^+...X^- + NaD \rightarrow W\text{-}NH_3^+...D^- + NaX \qquad [3]$$

In step [1], the first wool fiber, W, interacts with the second wool fiber, W', through the -NH₂ basic group which gains a hydrogen ion from the -COOH acidic group, creating the $-NH_3^+$ group which is positively charged, and the $-COO^-$ group which is negatively charged. The attractive force between the positive $-NH_3^+$ and the negative $-COO^-$ helps to bind the wool fibers to one another. In step [2], the wool is prepared for dyeing by the addition of HX, which is the general formula for an acid. If X = Cl, the formula would be HC1. The HX reacts with the –COO⁻ to restore the hydrogen ion, thus leaving a negative X⁻ ion to be attracted to the positive -NH₃⁺.This forces the two fibers apart at this point since they are no longer attracted to one another. In step [3], the dye D is added in the form of the sodium salt, NaD. The Na, which is positively charged as Na⁺ is attracted to the X⁻ and removes it from the fiber as NaX. The dye, as D⁻ then replaces the X⁻ and is now electrostatically attracted to the wool *via* the -NH₃⁺ portion of the wool molecule.

B. Hydrogen Bonding. All molecules containing –OH, -NH and FH groups undergo hydrogen bonding. This is because O, N and F are "electron hogs," that is, they strongly attract the electrons that constitute the chemical bond between themselves and the hydrogen. Thus the hydrogen will be partially stripped of electrons and will carry a partial positive charge. O, N and F all have unshared pairs of valence electrons, sometimes called "lone pairs," which will be attracted to the partially charged hydrogen on a neighboring molecule or a neighboring group in the same molecule. This attractive force is known as a "hydrogen bond." It can be illustrated thus:

$$-O\text{-}H + :O\text{-} \rightarrow -O\text{-}H....:O\text{-} \qquad [4]$$

The .... represents the hydrogen bond. The -O-H is called the donor group and the :O- is the acceptor group. If a fiber contains a donor group and a dye contains an acceptor group, mutual interaction between the two will fix the dye on the substrate. It is generally held that this is the chief process whereby wool, silk and the synthetic fibers affix their dyes, but cellulosic fibers use another mechanism since they have such a great affinity for water that it would be difficult for any dye to displace them.

C. Van der Waals Forces. Electrons of one atom may be very weakly and temporarily attracted to the nucleus of an adjacent atom. This force of attraction then sets up a partial, but temporary, charge on the overall atom. Such forces, known as van der Waals (after Dutch physicist J.D. van der Waals, 1837-1923) forces, are always present between dye and textile. Some authorities believe that forces of this nature are the most important factor in the absorption of dye molecules to a fiber.

D. Covalent Linkages. These linkages are due to the formation of covalent bonds between dye molecules and fibers. If a reactive group on a dye molecule reacts with some functional group on a fiber, the result could be condensation of the two molecules by a mutual sharing of electrons. For example, a chloride ion on a dye could interact with an -O-H group on a cellulose fiber in the following manner:

$$Dye\text{-}Cl + H\text{-}O\text{-}Cellulose \rightarrow Dye\text{-}O\text{-}Cellulose + HCl \qquad [5]$$

The dye is now bound to the cellulose by a covalent linkage Dye-O- .

17.4 Fastness Properties of Dyestuffs

A chemical substance requires more than just coloration to fit it for usage as a dye. For example, azobenzene is a red compound, but it is not a dye. A dye needs to be "fast," that is, stable, to wet treatments and to light. Washing is the principal wet treatment to which a dye is likely to be subjected once it is fixed on a fiber. Therefore, colorfastness on washing in water or dry cleaning with an organic solvent is a necessity for most textile dyes in today's competitive markets and practically all manufacturers in this country use washfast dyes. The severity of the test will depend upon the treatment the fabric will receive. For example, cotton may be placed in boiling water, whereas nylon may be washed in warm water and wool in cold water. On the other hand, wools are generally dry-cleaned, whereas synthetics are washed.

In order to enhance wet-fastness, a reactive group on a dye molecule is generally required to enable it to form a chemical bond with the fiber in the manner described in the previous section. Such groups often increase the conjugation of the conjugated system of the dye molecule and also serve as anchoring groups for the dye. Increased conjugation leads to intensification of color, so these groups can be rightly called auxochromes. Examples of some of these auxochromes are -O-H (phenolic group, attached to the benzene ring), $-NH_2$ (amino group), -COOH (carboxylic acid group) and $-SO_2OH$ (sulfonic acid group).

Dyes must also be resistant if exposed to perspiration, bleaching agents, dilute acids and dilute bases. Where such resistance is required, standardized tests are carried out. Such a test may be done simply by placing a drop of acid, base or bleaching agent on the fabric and visually assessing the appearance.

Dyed textiles are also routinely tested for lightfastness by exposing them to natural or artificial light for specific periods to rate them for color fading. The severity of these tests is also determined by the use to which the dyed textile will be put.

17.5 Usage (Application) Classification of Dyes

Dyes may be classified according to chemical structure or according to method of application. In each type of classification, there are several different schemes of arrangement, but no one scheme seems best. The following discussion classifies dyes on the basis of how they are used.

A. Basic Dyes. Basic dyes are salts which, when placed in water, dissolve to yield ions such that the positive ion, or cation, is the color-bearing portion of the molecule; the negative ion, or anion, is a simple ion such as chloride ion (Cl^-) or sulfate ion (SO_4^{2-}). The dissociation may be expressed by the following equation:

$$Dye\text{-}Cl \rightarrow Dye^+ + Cl^- \qquad [5]$$

The positive ion (the basic part of the dye molecule) possesses an affinity to fibers with negative groups. Basic dyes are used chiefly to dye Orlon and yield bright shades with excellent fastness properties. Basic dyes are also used to color protein fibers, wool, silk, nylon, and polyacrylics.

B. Acid Dyes. The acid dyes are the chemical opposites of the basic dyes. In the case of an acid dye, the colored portion of the molecule carries the negative charge and is the anion. Acid dyes find their main application in the dyeing of wool, but they are also used on a variety of other substrates such as silk, polyamides, acrylics, paper and leather. When wool is immersed in an acid dyebath, it already contains negatively charged carboxyl groups ($-COO^-$) and positively charged amino groups ($-NH_3^+$). The negative groups attract hydrogen ions, H^+, which are present in excess in acid solution, forming $-COOH$ groups. The overall wool molecule is now positively charged (due to the $-NH_3^+$ groups), and in order to maintain electroneutrality, it attracts the negatively charged dye molecules. In this way, acid dyes adhere to the fiber by the mechanism of electrostatic attraction.

C. Sulfur Dyes. The sulfur colors are a group of complex sulfur-containing organic compounds which must undergo several chemical steps in order to be fixed to a textile fiber. You recall from our discussion in the previous chapter than when an atom or a molecule loses electrons, it becomes oxidized. Chemists call opposite process, that is, a gain of electrons, "reduction." Sulfur dyes must first be reduced in basic solution in order to render them water-soluble. They are subsequently air-oxidized on the fiber itself to yield the desired shade. The process is usually carried out in a sodium sulfide bath since sodium sulfide can act as both a base and a reducing agent. Sulfur dyes are used mainly on cellulosic fibers such as cotton, linen and rayon. They usually have muted hues and exhibit poor lightfastness to sunlight.

D. Azoic Dyes. These dyes are generally formed right on the fabric by a chemical coupling process. For example, the dye para red can be produced in the test tube by reacting two chemicals, β-naphthol and the diazonium salt of para-nitroaniline. The formulas of these two compounds need not concern us here. It is sufficient to know that para red forms when the two starting materials literally couple together to form a larger molecule. In order to produce para red on a fiber, the fiber is first "padded" with β-naphthol by passing the cloth through a solution of this chemical and then squeezing it almost dry. In another bath, the diazonium salt of para-nitroaniline is prepared. The cloth impregnated with the β-naphthol is then immersed in the bath containing the diazonium salt and the chemical reaction takes place right on the fabric to produce para red. Water-insoluble azoic dyes can be applied in this manner to cellulosic fibers to give shades of high standard of fastness to light and to wet treatments.

E. Vat Dyes. This large and important group of water-insoluble compounds all contain a $-C=O$ group which, on reduction, is converted to $-C-O-H$ to produce a colorless, or leuco, form of the dye which is insoluble in water but soluble in basic solution such as sodium hydroxide (lye). The leuco form forms the salt $-C-O^-....Na^+$; it is in this form that vat dyes can adhere to textile fibers. Air oxidation over a considerable period of time restores the original shade of the dye. The indigoid dyes such as indigo and dibromoindigo

(Tyrian purple), and the anthraquinoid dyes, such as alizarin, comprise this group. Their major application is in the dyeing and printing of cotton. They are outstanding for their fastness properties and are also used for dyeing silks, wool and fur.

F. Mordant Dyes. Many natural coloring matters such as madder and logwood belong to the class of dyestuffs known as mordant dyes. These dyes cannot be applied to a fiber without the use of a substance called a mordant. The word "mordant" is derived from the Latin word "mordere," meaning "to bite," and refers to any substance which allows a dye to "bite into" or adhere to a fabric. The mordant acts as an intermediary which can form a bond with the fiber and with a dye. Mordants usually consist of metallic salts such as alum (potassium aluminum sulfate), chrome (potassium dichromate), iron (iron(II) sulfate), tin (tin(II) chloride) and copper (copper(II) sulfate), but other mordanting agents such as acetic acid, tannic acid, caustic soda (sodium hydroxide), lime, ammonia and tartaric acid are occasionally used as well. The mordant dyes must have at least one structural feature in their molecules which allows them to interact with and hold a metal ion. The interaction between the dye and the metal ion actually produces a complex ion of the type we discussed when examining colors produced by transition metal ions. Thus, the dye acts as a ligand with respect to the metal ion mordant. Since different mordants will produce different complex ions with the same dye, different colors are also produced. For example, when a chrome mordant is used on wool, dyeing with madder produces a garnet red shade. However, if alum is used as the mordant, a shade known as lacquer red results. Some natural dye sources and the colors produced with various mordants on several different substrates are listed in Table 17.2.

G. Disperse Acetate Dyes. This class of dyes is only very slightly soluble in water and almost all members of this group

TABLE 17.2
NATURAL MORDANT DYES FOR DYEING ANIMAL FIBERS

Source	Mordant	Substrate	Color Obtained
Black Oak (Quercus velutina)	Alum	Wool	Buff
	Chrome	Wool	Gold
	Tin	Silk	Orange
Bracken (Pteridium aquilinum)	Alum	Wool or Silk	Yellow-green
	Iron	Silk	Gray
Goldenrod (Solidage species)	Alum	Wool	Yellowish tan
	Chrome	Wool	Old gold
Marigold (Tagetes varieties)	Alum	Wool	Yellow
	Alum	Silk	Old gold
	Chrome-acetic acid	Wool	Buff
Sumac (Rhus glabra)	Alum	Wool	Yellowish tan
	Iron	Wool	Gray
Cutch (Acacia catechu)	Iron-chrome	Wool	Brown

contain amino or amino substituted groups. They are used for dyeing cellulose acetate, nylon, polyesters and polyacrylonitriles, as well as for the surface dyeing of plastics. The mechanism of fixation of dye to fiber is not clear, but it may be simple penetration of the fiber interstices by the dye molecules.

17.6 Selected Readings

Garfield, S. *Mauve: How One Man Invented a Color That Changed the World.* W.W. Norton: New York, 2001.

Greenfield, A.B. *A Perfect Red: Empire, Espionage, and the Quest for the Color of Desire.* Harper Perennial: New York, 2005.

Liles, J.N. *The Art and Craft of Natural Dyeing.* University of Tennessee Press: Knoxville, TN, 1990.

Orna, M.V. *The Chemical History of Color.* Springer-Verlag: Heidelberg, 2013.

Sequin-Frey, M. The Chemistry of Plant and Animal Dyes. *Journal of Chemical Education* **1981**, *58*, 301-305.

Weigle, P. *Ancient Dyes for Modern Weavers.* Watson-Guptill Publications: New York, 1974.

CHAPTER 18
ARTISTS' PIGMENTS AND COMMERCIAL PIGMENTS

18.1 Introduction

The second main class of colorants is pigments. Pigments may be defined as colorants which must be mixed with some kind of a binding medium, such as oil or egg yolk, before being applied to the substrate. Pigments must be insoluble in the binding medium in order to function properly.

Colorants have been used as pigments since prehistoric times, as evidenced by the cave paintings of Grotte Chauvet in France and Altamira in Spain, which have been shown to have been executed as much as 32,000 years ago. Chemical analyses at the caves in Altamira show that the pigments used were based mainly on iron and manganese oxides which are found in the earth as colored muds. These pigments provide the characteristic muted reds, yellows, buffs and blacks. In addition, carbon from burnt wood and yellow iron carbonate were used. The binders for these pigments were probably just water, with perhaps some bone marrow, animal fat or egg white mixed in. They survived the millennia in all their brilliance only because of their protected location.

In addition to the iron and manganese earth pigments mentioned above, the Egyptians, by 3,000 BCE, had succeeded in expanding the artists' palette to:

- Natural vermilion red (cinnabar, HgS)
- Red lead (lead oxide, Pb_3O_4)
- Malachite (basic copper carbonate, $2CuCO_3\ Cu(OH)_2$)
- Orpiment (arsenic sulfide, As_2S_3)

- Charcoal black and lampblack (carbon)
- Red madder (from the root of the perennial *Rubia tinctorum)*
- Egyptian blue (the first artificial pigment, formed as a glass by roasting sand and copper compounds together).

In the period dating from the Roman Empire, not as many new pigments were added. Among the new colors were:

- White lead (basic lead carbonate, $2PbCO_3$ $Pb(OH)_2$ (a pigment also employed in ancient China)
- Verdigris (mixed acetates of copper)
- Vegetable yellows and reds
- Smalt (blue cobalt glass)
- Natural ultramarine (processed from the semiprecious stone, lapis lazuli; it is a complex compound of soda, silica, alumina and sulfur).

The first modem synthetic pigment was Prussian blue, discovered in 1704. This discovery was followed by the synthesis of several additional inorganic pigments of cobalt, chromium and cadmium, so that within two centuries of the discovery of Prussian blue, the artist's palette had been expanded to more than twice its size. These syntheses made the artist, for the first time in history, independent of the vagaries of the natural pigments since the new pigments were relatively pure and reproducible in color and particle size from batch to batch. Table 18.1 is a fairly complete list of artists' pigments as they have been used through the centuries. We will discuss some of these pigments in more detail in later sections of this chapter.

In the nineteenth century, the artist also came to rely on packaged, prepared paints. Along with the decrease in self-prepared paints came the decrease in knowledge by the painter of the characteristics and properties of the pigments and paints. The artist of

TABLE 18.1
DATES FOR PIGMENT USE

Starting Date	Pigment	End Date
Before 1300	Asphaltum (Bistre)	
"	Azurite	1700s
"	Azurite + Lead-Tin Yellow	1720
"	Azurite + Yellow Ocher	1750
"	Blue Verditer	
"	Chalk	
"	Charcoal	
"	Cinnabar (Vermilion)	
"	Copper resinate	
"	Egyptian Blue	
"	Gamboge	
"	Green Earth (Terre Verte)	
"	Indigo	
"	Iron Earths	
"	Lampblack	
"	Lead-Tin Yellow	1750
"	Lead White (White Lead)	
"	Litharge	
"	Madder	
"	Malachite	1825
"	Massicot	
"	Minium (Red Lead)	
"	Orpiment	
"	Realgar	
"	Red Lakes	
"	Saffron	
"	Ultramarine (Natural)	
"	Verdigris	
1475	Smalt	~1825
1549	Cochineal	
1600s (Late)	Van Dyke Brown	
1700	Prussian Blue + Naples Yellow	
1700	Prussian Blue + Yellow Ocher	
1700	Prussian Blue	
1778	Scheele's Green ($CuHAsO_3$)	
1800	Barium sulfate	
1800	Chrome Yellow	
1800	Chrome Red	

TABLE 18.1
(continued)

Starting Date	Pigment
1800	Prussian Blue + Chrome Green
1800	Cerulean Blue
1802	Cobalt Blue
1809	Barium chromate
1809	Zinc Yellow
1814	Emerald Green
1817	Cadmium sulfide
1820	Chrome Green (Prussian Blue + Chrome Yellow)
1824	Ultramarine (Synthetic)
1825	Zinc oxide
1825	Chrome Red ($PbCrO_4.Pb(OH)_2$)
1825	Viridian ($Cr_2O_3.2H_2O$)
1825	Cobalt Green
1825	Cadmium Yellow
1826	Alizarin (Natural)
1842	Antimony Vermilion (Sb_2S_3)
1850	Prussian Blue + Cadmium Yellow
1850	Cobalt Blue + Naples Yellow
1850	Cobalt Blue + Cadmium Yellow
1850	Cobalt Yellow
1856	Coal-Tar Colors (Mauveine, etc.)
1859	Cobalt Violet (Cobalt phosphate)
1862	Chromium oxide (Cr_2O_3)
1868	Alizarin (Synthetic)
1868	Manganese Violet
1874	Lithopone ($ZnS + BaSO_4$)
1886	Aluminum Powder
1900	Barium sulfate
1910	Cadmium Red + Barium sulfate
1916	Titanium White (TiO_2)
1920	Antimony White (Sb_2O_3)
1926	Cadmium Red + Barium sulfate
1927	Cadmium Yellow + Barium sulfate
1930	Molybdate Orange
1935	Manganese Blue
1935	Phthalocyanine Blue

today must rely upon the color manufacturer's knowledge and techniques rather than upon his/her own expertise in the field of pigment behavior.

The tremendous range of reliable colors available today has been a privilege of the artist for only a relatively short period of time. The effect on modern painting is evident in the burst of experiment-ation and variation observable in the museums of modern art. Some of the credit for the difference between the art of today and of the past must go to the extended artists' palette.

All pigments are applied through the use of media such as oil, water, polymer resins or volatile liquids in which the pigment is insoluble. Unlike dyes, pigments are applied as a surface coating to wood, paper, canvas or other support. Dyes, on the other hand, must permeate their substrate and must be soluble during some part of their manufacture. The one current exception to pigmentation as a surface phenomenon occurs when pigments are incorporated into a plastic melt and color the whole mass, even though insoluble. When the plastic solidifies, the pigment particles are trapped in the mass.

Insoluble dyes may be used as pigments provided that their physical properties satisfy the strict requirements for a good pigment; then they are called toners. Soluble dyes must be specially treated to be used as pigments. They are precipitated onto inert, semi-opaque materials which support them. The supporting materials are called "lake bases" and they firmly hold the dye to form a "lake." Aluminum hydroxide, $Al(OH)_3$, is used as the lake base when a transparent color is desired as in printing inks. Barium sulfate, $BaSO_4$, is used as the lake base for opaque pigments.

18.2 Physical Properties of a Pigment

The physical properties of the pigments used in the fine arts must meet very stringent requirements. They are listed as follows:

A. Lightfastness (Colorfastness). The most important property of an art pigment is colorfastness. The pigment must retain its color under the attack of sunlight, artificial light and atmospheric chemical attack. Light is the most persistent enemy in the museum atmosphere. Although indoor lighting is only about 1/100 the intensity of sunlight, its effect is cumulative so that, over many years, a tremendous amount of exposure accrues.

B. Stability to Atmosphere. A pigment must be chemically stable and inert to the chemicals it may encounter. Oxygen and pollutants in the air can damage pigments. Sometimes, the medium forms a protective layer over the pigments as in oil and acrylic paints, thus protecting the pigment from environmental attack.

C. Media Adsorption. Media absorption refers to the fact that some types of pigment, because of their chemical composition, adsorb more oil (or water or acrylic solvent) when compounded than do others. Another factor which determines the amount of medium adsorbed is the amount of surface area of the pigment. The more finely powdered a pigment is, the greater the amount of surface, and the more oil or water is adsorbed. Particle size can be carefully controlled during pigment manufacture.

Oil, which acts as the vehicle for carrying a pigment, is adsorbed onto the surface of pigment particles by replacing the air that is already there. When the pigment has adsorbed as much oil as it can, the crumbly mixture becomes a paste because the oil helps the pigment particles to slide past one another. The amount of oil adsorbed by a given weight of a specific pigment can be used as a measure of its total surface area.

D. Opacity. Opacity is generally desirable in pigments since this increases tinctorial power. When more transparency is desired, as in water colors, extenders are added rather than switching to a less opaque pigment.

E. Bleeding. Bleeding refers to the tendency of a pigment to become soluble in adjacent layers, causing it to spread out of its own layer and to migrate to an adjacent one. Bleeding is a very undesirable property in a pigment.

The great majority of pigments in use today are synthetic rather than natural. Natural pigments are called native earths and are made of highly colored clays, soils and rocks. Permanent ochers, umbers, siennas, reds, yellows and blacks are produced from these materials. Red and black iron oxides and black manganese dioxide are the most important pigments among the native earths. To make some of these pigments redder, and occasionally more transparent, earths may be calcined or roasted in an oven. Most natural pigments require some chemical purification.

Inorganic synthetic pigments are made from ionic or mineral substances. They do not contain the undesirable impurities present in the natural materials. Many of these synthetic pigments are very permanent. The artificial earth colors are collectively called "Mars colors."

The next few sections of this chapter will discuss some of the major artists' pigments. Since whole books have been written on this subject, only some of the outstanding facts can be presented in this brief space.

18.3 Pigments from Chemical Elements

Perhaps the most common chemical element that yields pigments for commercial use is carbon. Carbon from various sources and in various forms is the chief chemical constituent of such pigments as the lampblacks, carbon blacks and graphite. In fact, virtually every artists' black pigment with the exception of Mars black (Fe_3O_4) is a form of elemental carbon. The carbon in all carbon blacks is the same and is exceedingly inert; differences between the various carbon blacks come from impurities and particle size differences.

The carbon blacks are produced by the incomplete combustion of natural gas according to the following equation:

$$CH_4 + O_2 \rightarrow C(s) + 2H_2O \qquad [1]$$

CH_4, methane, which comprises about 85% of all natural gases, combines with a limited amount of oxygen to form solid carbon, $C(s)$, and water. Other ingredients of natural gas may be ethane, C_2H_6, and propane, C_3H_8, and incomplete combustion of ethane and propane also produces $C(s)$.

There are many grades of carbon blacks which depend upon their "jetness," a property which is a function of dispersion of the carbon particles as well as their size. They are used for making solid color black goods of every kind and they are absolutely lightfast. A related group of black pigments, the lampblacks, are produced by burning creosote or other oils rich in hydrocarbons in a regulated amount of air in cast-steel pans housed in firebrick furnaces. The result is a pigment of much lower gloss than the carbon blacks; therefore, lampblacks are used in semigloss and eggshell finishes.

Another form of elemental carbon which occurs in nature in nearly pure form is graphite. In this form, carbon atoms occur at the corners of regular hexagons arranged in flat sheets (very nearly like old-fashioned bathroom tile), and each carbon atom has one unbonded, mobile electron. Because flat graphite sheets can easily slip over one another, graphite is very slippery. "Lead" pencils are made out of graphite mixed with varying amounts of clay to increase hardness. Very soft "leads" have little or no clay. When the "lead" is pressed onto the paper, the flat graphite particles layer themselves parallel to the surface to yield a slight gloss. Metallic lead has the same drawing properties as graphite and was once used for pencils. Fortunately, considering its poisonous character, it is no longer put to that use. Although graphite is a permanent color, it is not used in liquid paints.

The boneblacks, also known as animal black and ivory black, are usually discussed in the context of carbon black pigments but they are not, strictly speaking, chiefly carbon in composition. Most boneblacks contain less than 25% carbon and are mainly composed of a fine residual ash which is the result of charring bones. The grade of boneblack is highly dependent on the type of bone selected as the starting material.

Other elements which form pigments are metallic in character and yield pigments with a characteristic metallic color. Some examples are powdered aluminum, copper, zinc, tin, silver and gold. Bronze powders consist principally of a mixture of copper and zinc and produce a range of colors from red-gold with low zinc content, to pale gold, with a zinc content of ~30%.

18.4 The White Pigments

The white pigments are among the most important in the artist's palette. The three major white pigments in use today are white lead, $Pb(OH)_2.2PbCO_3$, zinc white, ZnO, and titanium white, TiO_2. We shall discuss each of these in turn.

Basic lead carbonate, known to artists as white lead, is a pigment which dates back over 2,000 years to the early Greek artists. It has always been the principal pigment in oil painting since its properties, when mixed with oil, are almost ideal. White lead is sold in this country for oil paints as Flake White. At one time, Cremnitz White was a special form of Flake White made from lead carbonate imported from Austria. Today its market is limited to people who insist on pure basic lead carbonate free from the additives normally incorporated into Flake White to improve its properties.

The old-fashioned process of making lead white by the Dutch pot process is still in use. Sheets of lead are placed into ceramic tanks containing very strong vinegar (28% acetic acid) and then covered with horse manure. The manure ferments to produce heat and carbon dioxide, CO_2 At 70°C, lead acetate forms and reacts with the CO_2, and water vapor to form $Pb(OH)_2.PbCO_3$. The entire process takes about three months. The modern chamber process uses the same chemical reactions but differs in spatial arrangement, the addition of CO_2, and the use of fuel to produce heat.

White lead is a superior pigment in oils because it has very low oil adsorption, reacts with oil to produce a buttery paint with excellent brushing qualities, has excellent opacity (hiding power), pleasing tone and is quite permanent. The reaction of the lead with oil forms a lead soap which contributes to the elasticity of the film. It dries more rapidly than the other white oil paints and forms a tough, durable flexible film.

White lead has two serious defects: it is highly toxic, and it may discolor on exposure to sulfur compounds. When eaten or inhaled, lead is a poison. Lead pigment should never be used in pastels, water color, fresco or any other water medium. It is relatively safe to use in oils because the pigment is enclosed in and chemically bound to the tough dry film formed by the dried oil. Because of its toxicity, white lead is banned from use in house paints, and for a while, was also forbidden for use in artists' pigments.

However, its qualities are so desirable that the government was persuaded to release it for artists' oil paints. It is quite safe for such use as long as one follows the simple rules of washing hands and fingernails thoroughly after use, and avoiding the pigment in powder form.

White lead can be affected by sulfur present in the atmosphere as hydrogen sulfide, H_2S:

$$Pb^{2+} + H_2S \rightarrow PbS + 2H^+ \qquad [2]$$

PbS, lead sulfide, is a black solid; therefore, its formation discolors any paint containing a lead pigment. H_2S may be present as a pollutant in industrial areas or in natural gas. However, when surrounded by the tough film of dried oil paint, white lead is protected from darkening. For this reason there seem to be few cases of darkening in well-protected oil paints.

Certain pigments such as cadmium reds and yellows and ultramarine blue are sulfides and could be incompatible with white lead if excess sulfide ion has not been removed. Therefore, only very high quality grades of these pigments should be used if they are to be placed in the same painting with white lead.

White lead is the best primer known because it forms the toughest of all oil paint films and is extremely durable. However, its opacity is less than that of titanium white, so the canvases primed with the latter sell better because they look whiter.

Flake White is supplied only in the professional line of oil colors and is not sold in cheaper lines of artists' colors.

Zinc white is the artist's name for zinc oxide, ZnO. The best grades are the Florence French Process zinc oxides, which are about 99% pure. Zinc white has two advantages over Flake White: it is non-toxic and does not discolor in the presence of sulfide since zinc sulfide is white. In oils, it is bluer and less opaque than Flake White, brushes out poorly, dries slowly, and forms a glossy but brittle film even though it is a reactive pigment. If opacity is desired in zinc oxide oils, a mixture of 50% zinc white and 50% titanium white or Flake White is recommended.

In all water mediums, zinc white is free of defects. Under the name of Chinese White, it is used in water colors.

Titanium white, TiO_2, is a dense, very opaque pigment with great hiding power. It was first marketed around 1920. It is absolutely inert and permanent; only hot sulfuric acid will dissolve it, and then only slowly. It is non-toxic, resistant to heat and does not yellow. It is the whitest of the white pigments. For these reasons, it is the most widely used industrial white pigment.

As an artists' pigment, titanium white has several disadvantages which have caused painters to prefer white lead. It is stringy when brushing and it dries very slowly. To make it less stringy, it is often mixed with 60% or more of barium sulfate or 50% zinc oxide, but this dilution also decreases its opacity.

Since both zinc white and titanium white are poor driers in oils, they must be well-dried before use if used as grounds. Titanium dioxide has a much finer body than white lead and therefore has high oil adsorption. In aqueous medium, titanium white is entirely satisfactory; it brushes well and is non-toxic and opaque.

There are two crystalline forms of titanium white, rutile and anatase. The rutile form chalks slowly and is the one used in artists' paints. Anatase chalks more rapidly and is used in house paints where chalking is desirable. Chalkiness causes softening of the paint film, so when a reactive pigment such as zinc white is mixed with titanium white, it counteracts the softening and makes the surface tougher.

Titanium white is the youngest of the white pigments, having been introduced to the paint industry about 1920. Its use rose rapidly in the 1930s and supply caught up to demand only in the 1950s. Since the date of its first use is so well-known, its presence in the painting can definitely date the painting as a post-1920 execution.

18.5. Colored Pigments from the Transition Elements

The elements in the first row of the transition elements (Period Four) that have colored compounds are vanadium, chromium, manganese, iron, cobalt, nickel and copper. With the exception of vanadium, these elements yield the most commonly occurring pigments. These are so numerous and have such different properties that they will be discussed in tabular form in Tables 18.2, 18.3, 18.4 and 18.5 on the following pages.

18.6. Pigments from Cadmium and Mercury

Cadmium and mercury, together with zinc, comprise the family IIB in the Periodic Table. Strictly speaking, they are not transition elements since their d-subshells are completely filled. Their compounds exhibit color due to charge transfer transitions and band transitions, not the d-d transitions associated with partially filled d-subshells.

Cadmium yellows, oranges and reds are produced by mixing cadmium sulfide, CdS, which is yellow, with varying proportions of red cadmium selenide, $CdSe$. The cadmium sulfide is generally coprecipitated with barium sulfate, $BaSO_4$, to form mixed crystals of CdS and $BaSO_4$; this mixture is called lithopone. The cadmium lithopone pigments are much more extensively used than the pure pigments because they are much less expensive. They are very bright, lightfast, and durable when they are the principal constituent of the pigment. Table 18.6 illustrates the various shades possible from these mixtures.

TABLE 18.2: COLORED PIGMENTS FROM CHROMIUM

Formula	Chemical Name	Color	Preparation and/or Properties	Trade Name
Cr_2O_3	Chromium(III) oxide	Dull olive green	Very stable, excellent lightfastness, unaffected by acid or alkali, can withstand temperatures of up to 1,000 °C. Non-bleeding, good hiding power	Chromium oxide green
$PbCrO_4$	Lead chromate	Medium yellow	When coprecipitated* with lead sulfate, $PbSO_4$ (white), the dark yellow of the chromate is modified to a lighter shade	Chrome yellow Light yellow Primrose yellow
$PbCrO_4 \cdot PbO$	Basic lead chromate	Light orange to red	Formed by precipitation of lead chromate from basic solution. Increasing amount of PbO leads to darker orange. Chrome red has same formula as chrome orange but differs in particle size, tetragonal plates and fine powder respectively	Chrome orange Chrome red
$PbCrO_4$ + Prussian blue**		Green	Mixture of chrome yellow and Prussian blue. Should not be used for permanent painting	Chrome green

*When compounds are coprecipitated, each crystal contains both compounds intimately mixed in a way which cannot be achieved by grinding them together.

**See Table 18.4 for the formula.

TABLE 18.3: COLORED PIGMENTS FROM MANGANESE

Formula	Chemical Name	Color	Preparation and/or Properties	Trade Name
$MnCO_3$	Manganese carbonate	Rose	Naturally occurring mineral, rhodocrosite	
$(NH_4)_2Mn_2(P_2O_7)_2$	Manganese ammonium pyrophosphate	Red to violet	Formed by fusion of manganese dioxide and ammonium phosphate; slightly soluble in water; lightfast; decomposed by acid and alkali	Manganese violet; also called mineral violet, permanent violet and Nurnberger violet
$BaMnO_4$	Barium manganate	Green-blue	Extremely stable pigment; unchanged by heat, acid or alkali	Manganese blue

TABLE 18.4: COLORED PIGMENTS FROM IRON

Formula	Chemical Name	Color	Preparation and/or Properties	Trade Name
$FeSO_4 \cdot 7H_2O$	Iron(II) sulfate heptahydrate	Green	Not a pigment but the starting material for the manufacture of other colored compounds of iron	Green vitriol Copperas
Fe_2O_3	Iron(III) oxide	Light reds	Heating of copperas at temperatures below 700 °C produces Fe_2O_3 with a disordered crystal lattice	Turkey reds
Fe_2O_3	Iron(III) oxide	Dark reds	Heating Fe_2O_3 at temperatures above 700 °C produces Fe_2O_3 with an orderly lattice	Indian reds
Fe_2O_3 + $CaSO_4$	40% iron(III) oxide + 60% calcium sulfate	Light red	Produced by heating copperas with lime, CaO. In oils, it produces a hard, brittle film	Venetian red
$Fe(NH_4)$- $Fe(CN)_6$	Iron(III) ammonium hexacyanoferrate	Greenish dark blue	First discovered in 1704 as an insoluble material with the formula $Fe_4[Fe(CN)_6]_3$, but now usually precipitated in the presence of an ammonium salt to yield compound shown at left, which has more satisfactory pigment properties. Very high tinctorial power, inexpensive, resistant to acid, lightfast, high oil adsorption; good hiding power; borderline permanence for fine arts	Prussian blue Milori blue Etc.

TABLE 18.5: COLORED PIGMENTS FROM SOME OTHER TRANSITION ELEMENTS

Formula	Chemical Name	Color	Preparation and/or Properties	Trade Name
$CoO.Al_2O_3$	Cobalt aluminate	Blue	Most important of the cobalt pigments; stable to heat and light; good for coloring ceramic glazes; can be used in all other painting techniques	Cobalt blue Thenard's blue
$C_{32}H_{16}N_8Cu$	Copper phthalocyanine	Deep blue	Commercially produced in 1936; very stable to heat, acid, alkali; good lightfastness; extremely high tinctorial strength	Monastral blue Thalo blue
$PbCrO_4$ + $PbSO_4$ + $PbMoO_4$	Mixture of chromate, sulfate and molybdate of lead(II)	Orange	High covering power and tinting power, but only moderately lightfast; used mainly in printing inks and commercial paints	Molybdate orange

The chief pigment of mercury is mercury(II) sulfide, HgS. It is a bright scarlet powder sometimes called Chinese vermilion or artificial cinnabar. It occurs naturally as the mineral cinnabar, and powdered cinnabar, known as vermilion, is one of the most ancient pigments known. Artificial vermilion, made by heating mercury and sulfur together, is also one of the earliest known synthetic pigments. When mixed with cadmium sulfide, it yields red to orange pigments which have been marketed under the trade name "Mercadium" pigments. Although vermilion has been known to retain its brilliance for centuries, it has also been known to darken upon direct exposure to sunlight. However, it has excellent body and hiding power, and is highly resistant to attack by acids and alkalis. It has often been used with white lead for flesh tints.

18.7. Other Inorganic Pigments

Only a few of the many other inorganic pigments will be mentioned here. A much more complete discussion is provided in the references at the end of this chapter.

Naples yellow is a heavy, semi-opaque pigment formed by heating (calcining) antimony(III) oxide, Sb_2O_3, with litharge, PbO. This process produces lead antimonate, $Pb_3(SbO_4)_2$. This compound is quite stable to heat, acid and alkali, but it turns to a permanent dark brown when heated to a very high temperatures. Since it contains lead, the precautions mentioned with respect to white lead apply.

Some compounds of arsenic such as orpiment, As_2S_3, which is yellow, and realgar, As_2S_2, which is orange-red, have been used as artists' pigments since ancient times. However, all compounds of arsenic are highly toxic and should be avoided.

Ultramarine blue is a crystalline sodium aluminum silicate with polysulfide groups built into the crystal lattice. It is the pres-

TABLE 18.6

**APPROXIMATE COMPOSITIONS OF CADMIUM
YELLOWS, ORANGES AND REDS IN LITHOPONE***

Color	CdS (%)	CdSe (%)	ZnS (%)	BaSO4 (%)
Golden	35	0	1	64
Lemon	33	0	2	65
Primrose	31	0	4	65
Orange	34	6	0	60
Orange-red	32	9	0	59
Light red	30	13	0	57
Medium light red	26	18	0	56
Medium red	21	25	0	54
Dark red	18	29	0	53
Maroon	16	33	0	51
Extra deep maroon	14	36	0	50

*Parker, D.H., *Principles of Surface Coating Technology*, Wiley-
Interscience, New York, 1965, pp. 120 and 134

sence of these polysulfide groups that give rise to the blue color.
Replacement of the sulfur by its close relative, selenium, yields a
blood-red color, and replacement with tellurium yields a yellow
color. Ultramarine is permanent and lightfast, and finds many uses
in art and industry. However, it works poorly in oils and is bleached
out by acids.

Natural ultramarine, derived from the semiprecious stone
lapis lazuli, was once more expensive than gold because the lapis
had to be transported overland from Afghanistan and the process of
purification was extremely laborious. However, the price dropped
dramatically when it was discovered how to synthesize this
material in the early part of the nineteenth century. Synthetic

ultramarine and natural ultramarine are chemically identical except for particle size, which is smaller in the former.

18.8. Organic Pigments

In past centuries, many organic materials such as cochineal, the madder lakes, sepia and saffron have been used as pigments. Since the middle of the 20th century, organic pigments covering every portion of the visible spectrum have been developed and intensive research has produced pigments that are not only brilliant, but also lightfast, heatfast and have high tinting power.

It is important to keep in mind that the same chemical compound can be used as either a pigment or a dye, depending upon the method of application. For this reason, *The Colour Index,* the standard work on the subject of colorants, classifies these materials according to use but also assigns a five-digit chemical identification number to the colorant; this enables identification of the compound even if the colorant appears in one section classified as a dye and in another classified as a pigment. For example, C.I. Vat Blue 14 is the same compound as C.I. Pigment Blue 22, so they both have the number 69810 assigned to them to indicate that they are chemically identical.

The high intensity of the modern organic pigments makes them very desirable for use in the artist's palette. However, they must also stand the test of permanence which is so characteristic of the inorganic pigments used in the fine arts. Consequently, each new organic pigment must be tested, in various paint formulations, by long exposure periods to the Florida sun. To survive this test, a pigment must demonstrate lightfastness equal to the inorganic pigments. Whether they will also survive the test of long exposure to atmospheric pollutants and oxidizing agents is unknown since these tests have yet to be made. Table 18.7 is a list of acceptable

TABLE 18.7
MAYER'S LIST* OF ACCEPTABLE ORGANIC
PIGMENTS

Most Acceptable**	Acceptable
Phthalocyanine blue	Acylamino yellow
Phthalocyanine green	Acrylamide maroon, medium
Hansa yellow	Carbazole diozazine purple
Green-gold (nickel azo yellow)	Flavanthrone yellow
Brominated anthranone orange	Parachlor nitraniline red
Anthrapyrimide yellow	Permanent carmine
Isoviolanthine violet	Permanent red FGR
Thioindigo red-violet B	Permanent yellow HR
Acridone red	Perylene maroon
Brominated anthanthrone orange	Pyranthrone scarlet
Quinacridone red (scarlet or yellow)	Quinacridone red (bluish)
	Quinacridone violet
	Red lake R (Copper complex)
	Thioindigo red-violet R-H
	Thioindigo red-violet X
	Thioindigo red-violet red
	Vat orange GR

*Mayer, R. The Artist's Handbook, 3rd Ed. Viking Press, New
York, 1970
**Fading comparable to inorganic pigments

organic pigments given by Ralph Mayer in *The Artist's Handbook*. Henry Levison has also published exposure data on organic pigments. The complete citation for this work is given in the section on selected readings.

Pigments, as colorants, are useless unless they are mixed with a binder to produce a paint, but pigments also behave differently in different binders. The next chapter will discuss the properties of paints.

18.9. Selected Readings

Bomford, D., *et al. Art in the Making: Impressionism;* The National Gallery: London, 1990.

Bomford, D., *et al. Art in the Making: Italian Painting Before 1400;* The National Gallery: London, 1989.

Feller, R.L., Ed. *Artists' Pigments: A Handbook of Their History and Characteristics, Vol. 1;* The National Gallery: Washington, DC, 1986.

Finlay, Victoria. *Color: A Natural History of the Palette*; Random House Trade Paperbacks: New York, 2003.

FitzHugh, Elisabeth West, Ed. *Artists' Pigments: A Handbook of Their History and Characteristics, Vol. 3;* Oxford University Press: New York, 1997.

Friedstein, H. A Short History of the Chemistry of Painting. *Journal of Chemical Education* **1981**, *58*, 290-295.

Gettens, R.J.; Stout, G.L. *Painting Materials: A Short Encyclopedia;* Dover Publications: New York, 1966.

Levison, H. *Artists' Pigments: Lightfastness Tests and Ratings;* Colorlab: Hallandale, FL, 1976.

Orna, M.V. Chemistry and Artists' Colors. *Journal of Chemieal Education* 1980, *57*, 256-269.

Patton, T.C., Ed. *The Pigment Handbook;* Wiley-Interscience: New York, 1973.

Roy, A., Ed. *Artists' Pigments: A Handbook of Their History and Characteristics, Vol. 2;* Oxford University Press: New York, 1993.

CHAPTER 19
PAINTS

19.1 The Composition of Paints

In order to be used for artistic purposes, a dry pigment must be incorporated into some sort of binder which will enable it to be spread evenly and smoothly. One method for achieving this is to blend the pigment with a solid so that the product rubs off with use to leave a colored streak; this is how pastel chalks and crayons function.

The more widespread usage is to support the pigment in a fluid; when this is done, the product is called a paint. In this chapter, the four major classes of artists' paints, oils, acrylics, tempera and water colors, will be examined

Artists' paints differ from house paints in several important respects. House paints are both protective and decorative; they protect a surface from abrasion, sun and water, and they decorate it by adding color, luster and the smoothing out of irregularities. Whereas an artists' paint demands permanence of color and a very stable surface under protected indoor conditions, house paints need to resist weathering and fading and may best achieve this by being able to gradually powder away even if this involves a color change (as long as the color change is uniform). House paints must be solvent resistant; artists' paints have no such requirement. House paints must appear smooth and uniform on the painted surface while artists' paints may emphasize technique through brush stroke. House paints are expected to last for about a decade, but artists' paints are expected to last indefinitely. Indeed, a ten year limit on the lifetime of artists' paints would wipe out the creative profession. Because of these differences, house paints contain many additives not desirable in artists' paints, and are compounded differently. They cannot be used as substitutes for artists' paints except for temporary commercial or educational needs.

The major ingredients of artists' paints are (1) the colorant and (2) the vehicle which supports and spreads the colorant. The colorant is the actual coloring material in dry form. The vehicle must contain a binder and a diluent. The binder anchors the pigment to the surface and prevents flaking; it may be a film as in oils, acrylics and tempera, or an adhesive as in water colors. Film binders (oil, alkyd, polyurethane, acrylic) also protect the colorant and supply gloss; they must be tough, flexible and durable, and they must form the film quickly. The diluent increases the fluidity of the paint to improve spreadability; it is usually a common solvent which is inexpensive and evaporates rapidly. The diluent for oil paints is turpentine; that for acrylics, tempera and water-colors is water. A paint without a colorant is called a varnish. Small amounts of additives may be added to paints to improve certain charac-teristics, and white pigments called extenders may be added along with the colorant to confer additional desirable properties or merely to decrease the cost of the paint. Figure 19.1 illustrates the relationships of these components to one another.

19.2 The Controlling Variables

When a pigment is mixed with a vehicle in paint manufacture, it forms a special kind of mixture known as a dispersion. Depending upon the pigment particle size, the dispersion may be a colloidal dispersion or a suspension. The relationship between solutions and dispersions is summarized in Table 19.1. The differences between these three types of mixtures are due entirely

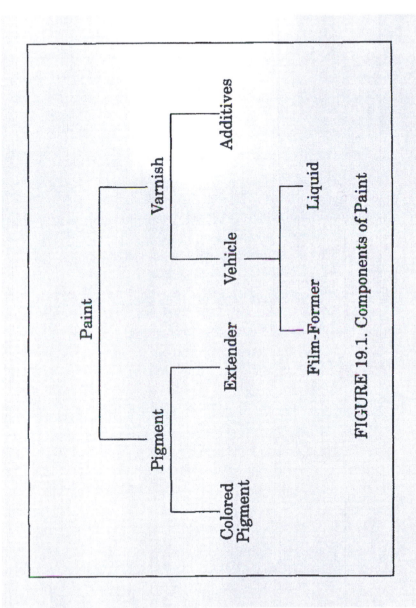

FIGURE 19.1. Components of Paint

to particle size and do not depend upon the nature of the solid material.

In order for a paint to be effective in hiding the surface beneath it, it must be opaque. True solutions are transparent, whereas dispersions have varying degrees of opacity, depending upon the light-scattering power of the dispersed solid. The higher the light-scattering power of a pigment, the greater its opacity because a high degree of light-scattering allows only little transmittance of light.

The light-scattering power, and therefore the hiding power, of a pigment depends upon two factors, namely, the pigment particle size, and the refractive index of the pigment relative to the medium in which it is dispersed.

A. Particle Size. A pane of glass is normally transparent because most of the light striking the glass is transmitted, and very little is reflected, scattered or absorbed. If a series of random scratches are made on the glass, the pane becomes less transparent because some of the light is scattered at the scratch interfaces. If the pane is shattered, still more light is scattered, and finally, if the broken pieces are ground up, so much light is scattered by the glass that a once transparent object is now opaque. A gradual decrease in particle size has increased the scattering power, and therefore the hiding power, of the glass.

The effect of changing the particle size can also be observed in materials which absorb light. If a green pigment, like malachite, is crushed and graded for size into the ranges 40,000-50,000 nanometers (nm), 20,000- 25,000 nm and 10,000-12,000 nm, as the particle size decreases, the amount of light scattered increases relative to the amount of light absorbed. This effect makes the smaller particles appear lighter in color. In fact, some pigments can only be ground very coarsely, because too much grinding causes their colors to fade. Continued grinding of the pigment particles will make them smaller and smaller. If it were possible to

TABLE 19.1
TYPES OF MIXTURES FORMED WHEN A SOLID IS ADDED TO A LIQUID

Type of Mixture	Particle Size (nm)	Properties
Solution	< 1 nm	Clear, transparent mixture; dissolved particles do not settle out on standing; particles are uniformly distributed; particles cannot be seen with neither the light microscope nor the electron microscope
Dispersions: Colloids	1 nm – 1,000 nm	Turbid, non-transparent mixture; particles do not settle out on standing; particles cannot be seen with the light microscope but can be seen with the electron microscope
Suspension	> 1,000 nm	Turbid, non-transparent mixture; particles settle out on standing unless coated with protective colloidal particles: particles can be seen with the light microscope

continue grinding the malachite particles until they were in the 250 nm range, maximum scattering of light would be observed, but further decrease in particle size would cause the degree of scattering to fall off rapidly. In general, the scattering power of pigment particles reaches a maximum when the particle size approaches half the wavelength of the incident light but falls off precipitately on either side of the maximum. This behavior is illustrated in Figure 19.2.

Since the wavelengths of visible light range from about 400 to 700 nm, pigment particles ranging in size from about 200 to 350 nm are the most efficient scatterers of visible light. However, it is not possible to achieve these very small particle sizes either by grinding or in the manufacture of a pigment, so both artist and decorator must be content with less efficient light-scatterers. Another effect of large pigment particle size is the fact that most paints fall into the category of suspensions, the solid particles of which may be expected to settle out on standing unless coated with protective colloids.

B. Refractive Index. The light-bending power of a medium is measured by its refractive index which is defined as the velocity of light in air or vacuum divided by the velocity of light in the medium. When light enters paint films of density greater than air, the higher the refractive index of the film, the greater the bending of the light. If pigment particles are suspended in the medium and have very nearly the same refractive index as the medium, then the pigment will bend the light to the same degree as the medium and very little light will be scattered. Therefore, the pigment-medium system will be transparent. If, on the other hand, the pigment particles have refractive indices very much greater or less than the medium, the light will be bent at very different angles by both pigment and medium and a great deal of light will be scattered. (See Figure 11.7 as a review of this behavior.)

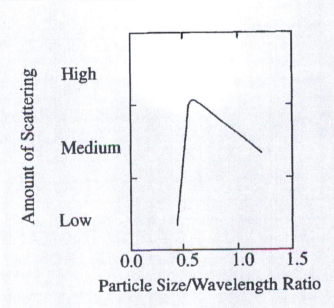

FIGURE 19.2. Scattering as a Function of
Particle Size for Titanium White

Such a pigment-medium system will be quite opaque. Thus, the light scattering power of a pigment will depend upon the difference between its own refractive index and that of the medium.

Tables 19.2 and 19.3 list the refractive index values of some representative pigments and vehicles in order of increasing refractive index. A glance at these tables shows that chalk, with a refractive index range of 1.5-1.64, when mixed with linseed oil, refractive index (symbolized by η, the Greek letter, eta) = 1.484, will exhibit a refractive index difference of about 0.02 to 0.16, which is a very small difference. Therefore, chalk scatters very little light when suspended in linseed oil and is so transparent that it is not even classified as a pigment. On the other hand, when chalk is suspended in a water-glue medium and spread out to dry, it is very effective as a pigment because after the water dries, it is the chalk-air interface with a refractive index difference of 0.5 to 0.64 (η for air = 1.00) which determines the light-scattering power. Thus, "whitewash," when properly used, can be very opaque. For the same reason, titanium white, TiO_2, with a refractive index of 2.5-2.6, is a very effective light-scatterer in any medium and is quite opaque. A little demonstration of this property can be made with the various liquid correction fluids on the market today. These products are nothing more than a suspension of TiO_2 in a volatile solvent. When applied to a piece of paper, the solvent quickly evaporates. The resulting white blotch reflects light almost as well as the paper when viewed in reflected light, but if the paper is held up to the light, no light is transmitted by the blotch and the area appears black in transmitted light. This simple procedure can demonstrate very effectively the hiding capacity of high refractive index pigments.

In addition to scattering power, the other two properties used to judge the desirability of artists' pigments are insolubility and stability. If a pigment has high light-scattering power, is insoluble in the medium in which it is to be used, and is very stable to heat,

TABLE 19.2

REFRACTIVE INDICES OF SOME ARTISTS' PIGMENTS

Name of Pigment	Refractive Index (η)
Smalt	1.49-1.52
Natural ultramarine	1.50
Aluminum hydrate	1.50-1.60
Chalk	1.50-1.64
Gypsum	1.53-1.62
Prussian blue	1.56
Vandyke brown	1.62-1.69
Malachite	1.65-1.88
Indian yellow	1.67
Azurite	1.73-1.84
Cobalt blue	1.74
Cerulean blue	1.84
Raw sienna	1.87-2.17
White lead	1.94-2.09
Zinc white	2.0
Naples yellow	2.01-2.28
Cadmium yellow	2.35-2.48
Orpiment	2.40-3.02
Realgar	2.46-2.61
Titanium white	2.50-2.60
Cadmium red	2.64-2.77
Vermilion	2.81-3.14

TABLE 19.3

REFRACTIVE INDICES OF SOME PIGMENT VEHICLES

Name of Vehicle	Refractive Index (η)
Water	1.330
Gum Arabic solution, 10%	1.344
Egg tempera	1.346
Linseed oil	1.484
Dammar	1.515
Shellac	1.516
Mastic	1.536

light and chemical attack, it is more than likely to be a suitable artists' pigment. Other properties of pigments such as effect on vehicle viscosity, oil adsorption and gloss are also criteria with which to judge desirability, but they are not nearly so important as the three named above.

Transparent pigments, called inert pigments or extenders, are often added to paints. The pigments which are used as extenders appear colorless and transparent when ground in oil or resins because their refractive indices are close to those of these vehicles. Extenders may be added to paints to increase film hardness or simply to reduce cost in the cheaper grades of paints. Student-grade colors are often loaded with considerable amounts of extenders as adulterants. This gives them desirable consistency (buttery, rather than stringy) but will eventually cause lowering in tone because of transparency. In addition, extenders will not mask the yellowing of oil in oil paints. When substituted for pigment, they weaken the tinting power of the paint. Titanium white, TiO_2, often contains at least 50% of the extender barium sulfate, $BaSO_4$, to make it less stringy, but as a consequence, it is also less opaque.

In aqueous medium, extenders may produce opaque brilliant white coatings since their refractive indices differ considerably from that of water.

Extenders in common use are aluminum hydroxide, $Al_2(OH)_6$, (sold as alumina hydrate), barium sulfate, $BaSO_4$ (sold as blanc fixe or as the less pure mineral barytes), calcium carbonate, $CaCO_3$ (precipitated chalk or whiting), and silicon dioxide, SiO_2 (sand or silica).

19.3 Oil Paint

The vehicle and binder used for oil paints are oils derived from various plant sources. The oil used most widely is linseed oil, which is extracted from the seeds of the flax plant, *Linum usitatissimum*. Other plant oils with more limited use are oiticica oil, safflower oil, perilla oil, poppy-seed oil and tung oil.

Before we can discuss the properties of oil as a pigment vehicle, it is necessary to look at its chemical constitution. Oils belong to that class of chemical compounds called esters, compounds that result when alcohols and acids combine with one another. The alcohol contained in oils is called glycerol, a sweetish, colorless, viscous liquid which, when mixed with lemon juice, has been used as a sore throat remedy. Its formula is $C_3H_5(OH)_3$, which gives one a partial idea of how the atoms are arranged with respect to one another. The extended structural formula of glycerol leaves no ambiguity in this regard:

$$
\begin{array}{c}
\text{H} \\
| \\
\text{H—C—OH} \\
| \\
\text{H—C—OH} \\
| \\
\text{H—C—OH} \\
| \\
\text{H}
\end{array}
$$

The -O-H group is characteristic of alcohols; glycerol contains three such groups and is called a triol. The acids which combine with glycerol to form oils vary from oil to oil. An organic acid has the general formula R-COOH, where the -COOH group is called the carboxylic acid group, and the R- is a general formula for a carbon chain. The three chief acids in linseed oil are oleic acid (22%), linoleic acid (17%) and linolenic acid (51%); each of these acids has an R- group containing seventeen carbons; their complete structural formulas are given in Table 19.4. These acids, with very long R-chains, are called fatty acids since they typically occur in fats and oils.

When glycerol combines with fatty acids to form fat or oil molecules, it will typically react with three of them because it has three -O-H groups. A typical oil molecule containing one molecule each of oleic acid, linoleic acid and linolenic acid attached to glycerol would look like the structure shown at the top of page 316.

However, an oil molecule may contain any combination of these fatty acids plus fatty acids that do not contain any double bonds. For example, an oil may contain two linoleic acids and one oleic acid, or another may contain three linolenic acids, *etc.*

TABLE 19.4
THE FATTY ACID COMPOSITION OF A TYPICAL LINSEED OIL

Name	%	Molecular Formula	Structure
Oleic	22	$C_{17}H_{33}COOH$	
Linoleic	17	$C_{17}H_{31}COOH$	
Linolenic	51	$C_{17}H_{29}COOH$	
Remainder	10		Long chains; no double bonds

```
     H  H  H  H  H  H  H  H  H  H  H  H  H  H  H  H  H  O        H
     |  |  |  |  |  |  |  |        |  |  |  |  |  |  |  ‖        |
 H—C—C—C—C—C—C—C—C—C=C—C—C—C—C—C—C—C—C—O—C—H
     |  |  |  |  |  |  |  |        |  |  |  |  |  |  |           |
     H  H  H  H  H  H  H  H        H  H  H  H  H  H  H           |
     H  H  H  H  H  H  H  H  H  H  H  H  H  H  H  H  H  O        |
     |  |  |  |  |     |        |  |  |  |  |  |  |  |  ‖        |
 H—C—C—C—C—C—C=C—C—C=C—C—C—C—C—C—C—C—C—O—C—H
     |  |  |  |  |        |        |  |  |  |  |  |  |           |
     H  H  H  H  H        H        H  H  H  H  H  H  H           |
     H  H  H  H  H  H  H  H  H  H  H  H  H  H  H  H  H  O        |
     |  |        |     |     |        |  |  |  |  |  |  |  ‖        |
 H—C—C—C=C—C—C=C—C—C=C—C—C—C—C—C—C—C—C—O—C—H
     |  |        |        |        |        |  |  |  |  |  |  |           |
     H  H        H        H        H        H  H  H  H  H  H           H
```

The most important chemical feature of an oil molecule is the presence of double bonds in the fatty acid portions of the molecule. These double bonds allow the oil to undergo polymerization in the same manner that ethylene polymerizes, as we saw in Chapter 16. If an oil molecule is symbolized by the general formula shown below,

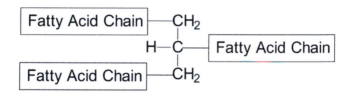

then polymerization between oil molecules can be illustrated by a scheme shown on the following page, where the heavy lines indicate new bonds formed.

As the oil polymerizes, larger and larger molecules are formed which create a film over the pigment particles. Eventually the film becomes one giant molecule. No one knows how long this process takes; it is possible that many centuries must elapse before polymerization is complete. The process takes a long time because the absorption of oxygen from the air, which is necessary to initiate the polymerization reaction, takes place very slowly, and because the subsequent reactions occur stepwise as a chain reaction. In a chain reaction, one reactant forms a second which reacts to form a third,

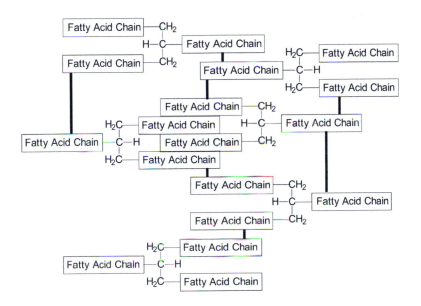

and so on. Some of the intermediate steps may take place very slowly and hold all the others up.

Exactly what happens as the polymerization chain reaction proceeds is known only in part. It is certainly known that the oxygen in the air is involved. The air, therefore, is a chemical reactant which is excluded from the paint by the tube. Since the tube is rolled or squeezed to extrude the paint, the contents are kept free of oxygen. No reaction occurs until the paint is spread in a thin film, presenting a large surface for reaction. Globs of paint take a long time to dry since the air takes a long time to permeate them.

Another factor that is essential for a satisfactory polymerization reaction is light. Absence of light as the reaction proceeds can cause darkening of the forming polymer film, a problem that can be corrected by exposure to light. However, direct sunlight causes cracking of the film because the reaction proceeds too quickly.

If an oil undergoes the polymerization process described above in a reasonable amount of time, that is, at about the same rate (or perhaps a bit slower) than the drying rate of linseed oil, the oil is called a "drying" oil. We must keep in mind that the "drying" process referred to here is not mere evaporation of water but a chemical reaction. A drying oil must contain numerous fatty acid chains which contain enough double bonds so that polymerization proceeds at a fast enough rate to form a film in a relatively short amount of time. In addition to linseed oil, several other drying oils were mentioned previously, namely, oiticica, safflower, poppy-seed, tung and perilla oils. If an oil undergoes the polymerization process so slowly that it takes a very long time for a film to form, it is called a semi-drying oil. Semi-drying oils contain fewer double bonds than drying oils. Some examples are cottonseed oil and sesame oil. Oils which contain so few double bonds that they do not form polymer films even after exposure to light and air for indefinitely long periods of time are called non-drying oils. Examples of non-drying oils are castor oil, coconut oil and olive oil.

The drying oils are of special interest to the painter, and particularly the vegetable oils obtained by the pressing or extraction of seeds or fruits. The most important of these vegetable oils is linseed oil, obtained from the flax plant. The best grade of linseed oil obtainable in this country is called cold-pressed oil. This is the kind usually sold in art supply stores; it may also be called refined, alkali-refined or varnish oil. Industrial oil, boiled linseed oil and other types of linseed oil contain impurities which cause yellowing and should not be used for artistic purposes. Certain oils are sold in which partial polymerization has been initiated; these include sun-thickened oil, stand oil and blown oil, all of which will dry faster because they are already partially polymerized. Blown oil tends to form a wrinkled film, but light stand oil and sun-thickened oil may be employed satisfactorily. Stand oil is relatively non-yellowing and dries to a smooth film free of brush marks, so it is recommended for glaze mediums, clear varnishes and emulsions.

Poppy-seed oil has less of a tendency than linseed oil to turn yellow and produces a buttery consistency more easily, but it dries slowly and tends to crack on aging.

The oil painting medium is easily manipulated and is adaptable to a wide range of effects. It accommodates pigments and glazes, and provides varying degrees of opacity as needed. Oil colors change very little on drying, a characteristic important to the artist. The paint film is flexible and slow-drying, thus allowing for corrections. In addition, oil paints have proven their durability over centuries of use.

Most painters feel that the advantages of oil outweigh the disadvantages. Oil medium darkens and yellows with time and can crack or flake. It is important to use proper techniques; violations, such as those perpetrated by Albert Pinkham Ryder, cause deterioration within a short time. For example, it is not advisable to apply a paint of low or medium oil content over a continuous film of high oil content since the upper layer will shrink more than the oilier layer upon polymerization. Such superimposition causes cracking of the film surface.

Commercially available oil paints must be carefully selected. In doing so, the painter should be aware of the fact that pigments may interact with the oil to influence its consistency, drying speed and flexibility. Among the pigments that are reactive are white lead and zinc white.

Some pigments have a tendency to absorb a great deal of oil, while others absorb less. To control the rate of drying and to adjust oil absorbency, manufacturers of artists' paints include additives in their oils which alter drying rates to keep the entire palette within some reasonable range. They also add stabilizers to keep the pigment in suspension and give the paint smooth-flowing characteristics. Since these additives may alter the stability of the paint film, reputable manufacturers try to keep additions to a minimum.

Oil paintings must eventually be varnished to protect the surface. Most varnishes contain a resin dissolved in a volatile solvent; when the varnish is applied to a surface, the solvent evaporates to leave the resin deposited as a clear film. Resins themselves are naturally occurring polymers which are exuded by certain coniferous trees. Dammar, mastic and copal are the chief resins used in varnishes. Although dammar and mastic resins dry through solvent evaporation, copal partially polymerizes on drying and thus forms a harder film. Dammar, however, is usually the preferred resin because it possesses greater elasticity, remains light in color, and does not promote bloom, a bluish film induced by trapped moisture.

The resins are soluble in turpentine and mineral spirits. They dissolve with difficulty and must be powdered to speed up dissolution. Copal must be melted (110 °C to 300 °C, depending upon the source) and re-solidified before it can be dissolved. Synthetic methacrylate polymer in petroleum solvent (or a blend with turpentine) provides a varnish which is less glossy than dammar.

Whatever the choice of material, paintings should be varnished within about three to six months of the completion of the work.

19.4 Acrylic Paint

The acrylic paints, as well as the alkyds, Glyptals and vinyl paints, all use as a binder an already polymerized material which is ground up and suspended in water. Of these, the acrylics are the most important and we shall confine our discussion to them, but all of the synthetic binders have properties similar to those of the acrylics.

There are two types of acrylic paints which have been marketed in the United States. The first, an acrylic material miscible with oil and turpentine, was introduced on the market in 1947 but never received widespread use. The second, a water emulsion, first appeared on the market in 1956, and became so popular that it is now the acrylic of choice.

The binder in all acrylic paints is a mixture of the polymers of methyl acrylate, ethyl acrylate and methyl methacrylate, with the formulas shown below. (The polymer of the latter is also called Lucite™ or Plexiglas™.)

Methyl acrylate   Ethyl acrylate   Methyl methacrylate

These molecules have double bonds which, as we have seen in the case of polymerization of ethylene and of oils, can open up to form single bonds that can link the molecules into long chains. Each of these polymer chains may be made up of 80,000 to 100,000 of the monomer units.

To make the acrylic vehicle, the polymer is ground up and suspended as a very fine powder in water. Since this mixture is not a solution, it is milky in appearance because of the very fine particles suspended in the aqueous phase. When a film of this suspension is applied to a surface, the water evaporates, causing the polymer molecules to gradually approach each other. Eventually, they coalesce to form a uniform, highly transparent, flexible, continuous film The film is semi-crystalline, which means that some of the polymer chains are lined up uniformly alongside each other to form the crystalline part of the film, and other chains are randomly arranged, forming the non-crystalline portion of the film. Figure 19.3 illustrates how the strands of a polymer can be arranged at random, in an orderly array, or can be random in part and orderly in part.

Simple water suspensions are difficult to apply as paints because they flow too easily over the surface. Thickeners must be added either to the polymer phase or to the solvent. Resins may be added as polymer thickeners; plasticizers such as dibutyl phthalate which make the film more flexible also help to make the paint more viscous. Thickening agents for the aqueous phase include sodium carboxymethyl cellulose. Surfactants are also added to keep the polymer particles from coalescing prematurely.

Acrylic resin was first synthesized by Otto Röhm (1876-1939) in 1901, but it did not gain popularity as a possible artists' medium until the 1960s. Although all brands of acrylic paints were formerly made from Rhoplex A34, a polymer manufactured by the Rohm and Haas Corporation, improved polymer formulations by other manufacturers have since come on the market to provide an array of choices for the artist in professional and student grades. These acrylic paint formulations contain various additives that may be incompatible with one another, and so it is not wise for the artist to mix different brands without testing for incompatibility first. For example, mixing paints of different brands may cause the mixture to curdle (clump out of the suspension). One way of checking for

 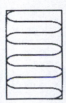 

Figure 19.3a	**Figure 19.3b**	**Figure 19.3c**
Random arrangement of a Polymer Strand	**Laminar arrangement of a Polymer Strand**	**A Polymer Strand partly ordered & partly random**

Figure 19.3. Possible Arrangements of a Polymer Strand

These states can be illustrated by using long strands of cooked spaghetti. Figure 19.3a can be simulated by tossing the strands into the air and letting them fall at random. Figure 19.3b can be reproduced by layering a strand in the manner illustrated. Figure 19.3c can be simulated by partially disturbing the order of the strand in Figure 19.3b. This model was suggested by Charles E. Carraher, Jr. in "What Are Polymers?" *Chemistry* **51(5), pp. 6-10 (1978)**

curdling is by mixing two well-thinned samples of paint on a strip of glass or plastic and observing the results.

A typical acrylic paint formula will include more or less equal weights of pigment, polymer (including up to 15% plasticizer) and water, with small amounts (less than 1%) of dispersant, surfactant, thickener and acidity controller.

Acrylic paints dry very rapidly, usually within thirty minutes, as the water evaporates off and the polymer coalesces to form a water-insoluble film. The dried film is soluble in certain organic solvents such as turpentine, acetone and xylene. The artist using acrylic paint must be careful to keep the brush wet so that it does not dry out to a hard block. It is best to use a palette such as glass or filled marble which can be scraped clean with a razor blade.

As with any medium, there are advantages and disadvantages to the use of acrylic paints. An acrylic film is durable and does not discolor. It may be easily overpainted, can be built up to any thickness, tolerates inert pigments and can be cleaned with a damp cloth. It is also quite elastic and maintains its flexibility over a long period of time. However, since oil paints tend to become brittle with time, they should not be used over acrylics. It appears to be safe to use acrylics over all water media colors.

In handling acrylics, they can be treated as water-miscible paints such as tempera since they share the characteristics of aqueous paints rather than oils. Since no volatile solvent is used, they are completely non-toxic. However, because they dry so rapidly, it is not easy to blend or gradate hues and tones nor can they be varied in opacity. Also, although acrylics produce brighter colors with the same pigments that are used in oil medium, acrylic colors look quite different after drying. Some pigments, such as lead white, cannot be used in acrylic medium.

Although water-based acrylic paints for artists were introduced in the early 1960s, their widespread use is only a little over 25 years old, so many problems have yet to be worked out. In the use of acrylics, it would be wise for the inexperienced artist to experiment with caution.

Some well-known artists who popularized the use of acrylics are Andy Warhol (1928-1987), Robert Motherwell (1915-1991) and David Hockney (b. 1937).

19.5 Tempera Paints

Before the advent of acrylic paints, tempera paints had been traditionally defined as paints which could be diluted with water, but which formed an insoluble film on drying. However, since acrylic paints also fall within this definition, a new one must be formulated for tempera paints. Tempera paints are oil-based paints which can be diluted with water and which form an insoluble film on drying.

All tempera paints use oil as the binder, but the oil is homogeneously mixed with water to form an emulsion. This oil-water emulsion is the vehicle in which the pigments are suspended to form the tempera paint. Emulsions are stable mixtures of two immiscible liquids interspersed uniformly in each other.

The mechanism by which tiny droplets of one liquid hang uniformly in another liquid without settling out depends upon the presence of an emulsifier or stabilizer. Usually, the emulsifier is a long-chain molecule which is partially ionic at one end, but composed of covalent bonds in the rest of the molecule. The partially ionic portion of the molecule can dissolve in water while the covalent portion dissolves in oil. In Figure 19.4, the emulsifier is represented as

where the pointed end is the partially ionic part of the molecule and the wavy line represents the long-chain covalent part. The

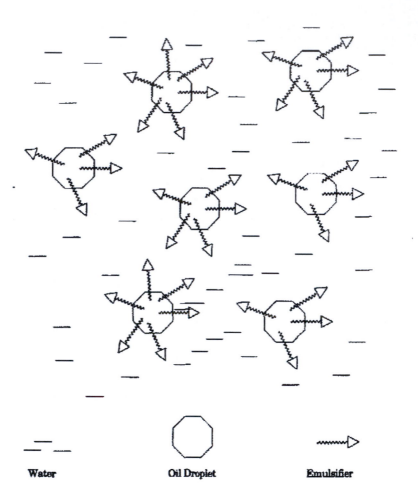

Water   Oil Droplet   Emulsifier

FIGURE 19.4. Model of Emulsification

partially ionic "heads" remain in the water phase while the covalent "tails" dissolve in oil droplets. In this way, the oil droplets are dispersed throughout the water. Soaps and detergents wash out organic dirt by this same "split personality" method.

The emulsifiers in common use in tempera paints are gum Arabic, glue and egg. Gums are water-soluble exudates of trees and shrubs and have the typical emulsifier structure; glues are proteins, which are condensation polymers of amino acids. Each of these emulsifying agents acts in the manner described above to mix the oil and water which comprise tempera paint vehicles. In the trade, tempera emulsifiers are called stabilizers. Emulsion vehicles used for tempera paints include oil-egg-water, varnish-gum Arabic-water, oil-glue-water, oil-egg-varnish-water, and other such combinations.

Emulsions based on glue and gums must contain a preservative such as 1-2% para-chloro-meta-cresol or 0.1% phenyl mercuric acetate because the emulsifiers are organic materials which can serve as food for funguses and bacteria. Ingestion of tempera paints should be avoided since mercury may be present.

The chief type of tempera paint used in earlier times was based on a natural emulsifier, egg yolk, which is still quite popular. Egg yolk is an excellent emulsifier since it is a water solution of albumin (a gum protein), lecithin and cholesterol (fat-like substances which are very efficient emulsifiers), and a non-drying egg oil. About 49% of the egg yolk is water.

When egg yolk is spread out in a thin layer and allowed to dry while exposed to daylight, the egg albumin coagulates in the same way that egg white solidifies when cooked. The egg yolk dries to a tough, permanent film which imparts an interesting quality to paints. The egg tempera medium allows the artist to slowly build layer upon layer of paint, each so thin that it is translucent. However, all egg yolk-oil emulsions do not have identical textures because egg yolks vary in the amounts of lecithin and cholesterol they contain. Lecithin stabilizes oil-in-water emulsions, such as salad dressings, while cholesterol stabilizes water-in-oil emulsions, such as butter. Since these actions oppose one another, the texture of an egg tempera paint will depend upon the lecithin/cholesterol ratio in the egg yolk used. The yellow in the egg yolk does not affect the pigment colors in egg tempera paints. Egg tempera may be preserved with commercial vinegar, a 5% acetic acid solution.

Tempera paintings in general are characterized by bright, crisp colors which, when dry, resemble their original dry state more than do oil colors. The dried paint film does not yellow or darken with age. Drying tends to be rapid since the bulk of the liquid is water. However, while colors can be over-painted, they are difficult to gradate or blend.

19.6 Water Colors, Gouache Paints and Poster Colors

Water colors are made of very finely ground pigments suspended in water, usually with gum Arabic as a binder. The particles of pigment become embedded in the paper and, like a stain, firmly adhere to it. Since water colors do not form an adhesive coating, peeling and cracking do not occur.

Water color paints contain essentially the same pigments used in oil paints. When used in water medium, they are left on the paper exposed to air. If protected from abrasion and damage to the paper, they are just as permanent as oils. However, pigments which are poisonous should not be used as water colors since there is no protective coating over them such as that which forms in oil paints.

Although water colors can be home-made, their manufacture is difficult primarily because of the very fine grinding that is needed. Hand-ground colors tend to be grainy and do not adhere well to a surface.

Gouache paints are opaque water colors. The pigments are ground with an increased proportion of paint to vehicle. Thus a paint film of some thickness forms and an actual layer of paint is created, in contrast to the thin wash or stain of water colors.

Gouache paints do not need to be as finely ground as do water color pigments, so they can be made much more easily. It pays for artists to grind their own gouache colors both for control of materials and for economy. The recipes are similar to those for water colors but may include up to 50% inert pigments such as chalk or *blanc fixe* in order to create desired opacity and texture. Prussian blue may require more inert pigment than most other colors. Only small amounts of gum solution are needed, perhaps only as little as 10-15 grams per 50 grams of total pigment. Precise paint recipes cannot be given because of variations in pigment preparation. Mayer *(The Artist's Handbook,* 3rd ed., pp. 308-311) has a number of recipes for gouache paints.

Poster colors are made up of inexpensive pigments, extenders or fillers to which a water-soluble binder, preservative, and sometimes a little glycerin have been added. They are impermanent, and thus suited for such uses as commercial displays and educational uses. It is especially important to check poster colors for poisonous ingredients such as lead, mercury or arsenic if they are to be used in schools. Organic colors should only be used when children are mature enough to avoid getting paint into the digestive or respiratory systems.

As a general practice, brushes should not be moistened with the mouth when working with water colors nor, for that matter, when working with any pigment materials. It is wise to follow simple rules of cleanliness whenever working with paints.

Some rules for safe handling of artists' materials will be given in Chapter 23 on "Art Hazards."

19.7 Selected Readings

Bentley, J.; Turner, G.P.A., *Introduction to Paint Chemistry and Principles of Paint Technology*, 4th ed.; Chapman & Hall: New York and London, 1998.

Gettens, R.J.; Stout, G.L. *Painting Materials: A Short Encyclopaedia;* Dover Books: New York, 1966. A comprehensive manual of pigments, dyes, and vehicles, with historical notes and references. A must-have "oldie" but "goodie."

Gottsegen, Mark *Painter's Handbook*, Revised and Expanded; Crown Publishing Group, New York,, 2006

Greenberg, B.R.; Patterson, D. *Art in Chemistry, Chemistry in Art*; Greenwood Publishing Group: Westport, CT, 2008, especially Chapters 2 and 3.

Chapter 20

Photography

**I am a camera with its shutter open,
quite passive, recording, not thinking…
Some day all this will have to be
developed, carefully printed, fixed.**

Christopher Isherwood

CHAPTER 20
PHOTOGRAPHY

20.1. A Brief History of Photography.

The need to record events and persons as well as abstract concepts is one that is very evident in the history of art. Between the earliest cave painter and the most accomplished portrait painter of the nineteenth century stands a long tradition of recording almost any subject matter conceivable. It is only natural that inventors should experiment with the means of producing highly realistic images and should devise ways of preserving them in permanent records. It was not long before the combination of these two processes in a new representational tool, photography, was developed and adopted by artists as an aid to art. Eventually, it grew into an art in its own right.

No technical process evolves by itself. Evolution in auxiliary, and sometimes in seemingly unrelated areas, is necessary for the final "coming together" of technical achievement. Photography was no exception to this. Although some of the basic principles of optics were known in ancient times and although the formation of images by pinholes was mentioned by Aristotle, it was not until the sixteenth century that the elements of photography began to coalesce. By that time, the *camera obscura* (popularized by the Neapolitan physician, Giambattista della Porta, 1535-1615) fitted with a lens was a popular entertainment device and sketching aid. By the end of the century, a practical knowledge of lenses, optics and image formation served to produce the microscope and the telescope. Thus, the essential ingredients of the modern photographic camera and the principles of image formation were well understood over four hundred years ago, but it took another three hundred years before anyone succeeded in making a permanent record of these optical images. This could only happen with the evolution of chemical science.

Over the course of the preceding 1,600 years, while the principles of optics were being worked out, the discipline of what came to be known as physics was dominant. As the chemical sciences evolved in the 18th and up to the mid-19th century, chemists, would-be chemists and entrepreneurs experimented with substances that seemed to be light-sensitive and thus able to react and eventually permanently record an image. Thus, the chemical sciences came to the fore and by the middle of the 19th century, more or less satisfactory methods of image recording were becoming routine. Continued experimentation improved these processes enormously throughout the 20th century, but by the late 1980s, it was becoming evident that digital photography, a method that bypasses chemical reactions, was in the ascendancy – and by the beginning of the 21st century, the field of photography was squarely back in the domain of physics. The chemical processes of photographic development, fixing, and printing were virtually a thing of the past – a niche market in specialty photography. However, it is well to look at the history of photography through all of these phases in order to understand how scientific understanding in many areas converged to give us this art form.

In order to render an optical image permanent, it is necessary that the light forming the image also have the power to produce a chemical change in the light-sensitive material medium. Silver compounds have been known since at least the thirteenth century, and it has long been known that the halides of silver, $AgCl$, $AgBr$, and AgI, are outstanding in their sensitivity to light. Around the beginning of the nineteenth century, Thomas Wedgwood (1771-1805), the son of the porcelain manufacturer, Josiah

Wedgwood (1730-1795) and Sir Humphry Davy (1778-1829), were among the first to produce images on paper impregnated with silver nitrate ($AgNO_3$) or silver chloride ($AgCl$), but neither discovered a way of preventing their papers from darkening upon exposure to light. It was not until 1826 that Nicephore Niepce (1765-1833), a retired businessman from Chalons-sur-Saône, succeeded in taking the first successful permanent photograph by coating a metal plate with asphaltum and, after exposure, by dissolving the unexposed and more soluble portions with aromatic solvents. This process proved to be quite impractical, and after Niepce's death in 1833, his partner, Louis J.M. Daguerre (1787-1851), turned to quite a different process. In essence, daguerreotyping consists of exposing a silvered copper plate to iodine fumes to form a surface layer of silver iodide (AgI). After exposure, the plate is then treated with mercury vapor which preferentially deposits itself on the exposed portion to produce a visible image. The plate is then washed with a sodium hyposulfate solution in which the unexposed portion of the plate is soluble, thus yielding a positive, lightfast, photographic image. This was the most popular process in the decades 1840-1860.

From this brief description, it is evident that the daguerreotype was clumsy, expensive, and quite hazardous. During its heyday, an Englishman named William Henry Fox Talbot (1800-1877) was carrying on experimentation which would eventually earn him the title "Inventor of Modern Photography," although not without some controversy. Fox Talbot built his photographic business on the shoulders of many predecessors including Thomas Wedgwood and Sir John Herschel (1792-1871). For example, although Fox Talbot is credited with the concept of the photographic negative, it was actually Herschel who had originally coined the terms "negative" and "positive" with reference to the photographic process.

20.2. Development of Modern Photography.

Throughout the nineteenth century, experimentation with light-sensitive chemicals in order to provide the photographic image continued. The most exploited of these chemicals were the silver halides, AgCl, AgBr and AgI. It was well known that these silver salts were not only very slightly soluble in water, but also quite light-sensitive. If one takes an aqueous solution of silver nitrate, $AgNO_3$, and mixes it with an aqueous solution of potassium bromide, KBr, a yellowish precipitate of silver bromide, AgBr, forms according to the following chemical equation:

$$AgNO_3(aq) + KBr(aq) \rightarrow AgBr(s) + KNO_3(aq) \, [1]$$

This precipitate darkens upon standing exposed to light, so it occurred to the early experimenters that its light-sensitive properties might be useful in forming the photographic image.

The first person to actively use AgBr was John F. Goddard (1797-1866) in 1840. He did it by exposing Daguerre's silver-copper coated plates to bromine vapors as well as iodine vapors, thus effectively carrying out two reactions:

$$2Ag + I_2 \rightarrow 2AgI \qquad\qquad [2]$$

$$2Ag + Br_2 \rightarrow 2AgBr \qquad\qquad [3]$$

Along with analogous reactions of copper with these halogens, inclusion of AgBr with the AgI greatly improved the daguerreotype process since photographic exposure time could be reduced to several seconds.

In the meantime, other persons such as Fox Talbot and J.B. Reade (d. 1870) were experimenting with the impregnation of paper with soluble silver halides and subsequent reaction with halide to produce the insoluble silver halides absorbed on the paper. A distinct improvement on this method was achieved by Frederick Scott Archer (1813-1857) in 1851, who succeeded in producing a relatively evenly distributed suspension of silver in collodion, a jelly-like substance that is made by dissolving guncotton in ether. However, such a photographic plate was quite inconvenient since it had to be exposed and developed while still wet, and therefore had to be made on the spot! This process gave rise to itinerant photographers who carried not only their cameras with them, but also the equipment and chemicals necessary to manufacture and process their film and, in addition, a portable darkroom in which to carry out all these processes! In 1871, a physician named Richard Leach Maddox (1816-1902) was able to eliminate this inconvenience by using gelatin as the suspended medium for the silver halide. He added silver nitrate and potassium bromide to the warmed gelatin mixture and allowed silver bromide to form according to equation [1]. He then coated glass plates with the mixture and allowed them to dry. At first, potassium nitrate, the other (and unwanted) product in equation [1] also crystallized out on this photographic emulsion (really a suspension), but this difficulty was remedied by allowing the gelatin mixture to solidify. It was then shredded under running water to dissolve out the water-soluble potassium nitrate, reheated to form a liquid, and then coated on the glass plates. Thus, the modern photographic emulsion was born: a mixture that remains, with some refinements, essentially the same to this day.

In the meantime, what of our inventor of modern photography, FoxTalbot? He had nothing to do with the aforementioned technical improvements, yet he walked away with all the laurels. Why?

Let us backtrack to about 1840. Around this time, Fox Talbot was attempting to increase the silver concentration of his impregnated paper in order to reduce exposure time, and he actually succeeded in getting the latter down to about twenty minutes. Then his lucky accident occurred. Remember that the sensitized silver halide-containing paper required numerous washings in silver nitrate to attain the required sensitivity. Fox Talbot heard that J.B.. Reade had succeeded in increasing the sensitivity of the paper still further by adding gallic acid, a mild reducing agent, to the last wash. He tried this procedure himself, but accidentally exposed some paper with insufficient gallic acid and produced a slightly visible image. When he tried to retrieve his error by placing the already exposed paper back into a silver nitrate bath in order to resensitize it, much to his astonishment, the original image appeared. A barely visible image due to exposure had been on the paper all along. It required only "development" by the mild reducing agent, gallic acid, to appear as a full-blown silver image. Thus Fox Talbot's discovery of the latent image, which now reduces the necessary exposure time to only a few seconds, marks the beginning of modern photography.

20.3. The Latent Image.

The chemical reaction by which a photographic film is exposed is simple:

$$2AgBr \rightarrow 2Ag + Br_2 \qquad [4]$$

This equation tells us that the silver halide emulsion (assuming that the silver halide is silver bromide) is induced by light (symbolized by hv) to become elemental silver atoms, Ag, and elemental bromine atoms, Br, which always pair up to produce bromine molecules, Br_2. Hence the need to represent the simultaneous reaction of the silver bromide, though simultaneity isn't necessarily so.

If we dissect this reaction, we can write two partial, or half, equations which depict what is happening to the silver and bromide separately:

$$Ag^+ \rightarrow Ag \qquad\qquad [5a]$$

Here, a positively charged silver ion from the silver bromide salt is becoming a neutral silver atom. Since it is becoming less positive, it must be becoming more negative by gaining a negative charge. To do this, the Ag^+ ion must gain an electron. Thus the completed half-equation is:

$$Ag^+ + e^- \rightarrow Ag \qquad\qquad [5b]$$

Note that the sum of the charges on each side of the equation must be equal. Chemically, we define any process in which the electrons are gained as reduction. Thus we see that, in this case, the silver ion has been reduced to a silver atom. (Remember that the gallic acid was a mild reducing agent. What do you think it did to Fox Talbot's paper?) However, for every reduction half-equation, there must be a corresponding half-equation that depicts a loss of electrons in order to supply the electrons gained during the reduction. Chemically, the process of losing electrons is termed oxidation, and thus, an oxidation half-equation must be written.

Before we do this, however, let us take a brief look at the structure of the silver halide emulsion in the photographic film. When the silver halide precipitates upon formation, it does so as crystals.

This is because silver ions (Ag⁺) are positive and the bromine ions (Br⁻) are negative. Therefore, they not only attract each other on a one to one basis, but each positively charged silver ion attracts as many negatively charged bromide ions as it can possibly fit around it, and bromide ion will do the same. This aggregate then begins to look something like Figure 20.1, but in a three-dimensional array with bromide ions sitting above and below the silver ions lying in the plane of the paper, and the silver ions sitting above and below the bromide ions. This array of ions is known as a crystal lattice. Note that the silver ions are quite small (though not drawn to scale) and are able to move somewhat in this relatively rigid lattice. However, the bromide ions are a bit too large to move about freely.

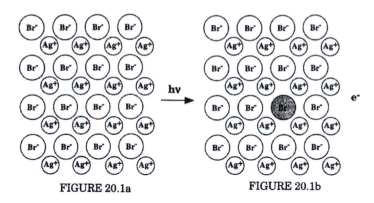

FIGURE 20.1a   FIGURE 20.1b

FIGURE 20.1. Production of a Positive Hole by a Photon (hv) in a Silver Bromide Lattice

Silver halide crystals occur as autonomous aggregates called "grains" in a photographic emulsion. The size of these grains can be carefully controlled in the film-making process. As we will see shortly, grain size can greatly affect the necessary exposure time.

When a photographic film is exposed to light, the interacting light can be thought of as little packets of energy called photons. A photon is symbolized by hν, but since it is not a material object like an atom or an ion, it can only be represented in a chemical equation by sitting on the arrow. It does not materially take part in the reaction, but supplies the energy necessary to induce the reaction. In that sense, it is consumed, or partially consumed, but this cannot be represented in a chemical equation. When a photon strikes a silver halide grain in a photographic emulsion, it sometimes "knocks out" or releases an electron from a bromide ion in the following manner:

$$Br^- \rightarrow Br + e^- \qquad \text{[6a]}$$

Knowing your rules of equation writing, you can also represent this reaction as

$$Br^- - e^- \rightarrow Br \qquad \text{[6b]}$$

indicating that bromide ion has lost an electron. This is therefore an oxidation half-equation, and the oxidation counterpart to equation (5b). However, chemical convention demands that we never depict substances as being lost. Hence, the first representation of this equation is the correct one. Note that once again, the sum of the charges on each side of the equation is equal.

Now, if the bromide ion becomes a bromine atom by the action of a photon, something happens to the charge balance in the crystal lattice of Figure 20.1. The new situation is depicted in Figure 20.2, where one of the bromines is now neutral, thus giving an

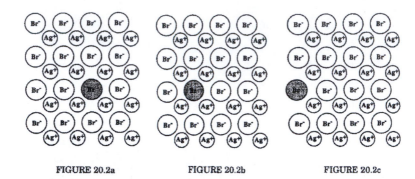

FIGURE 20.2a   FIGURE 20.2b   FIGURE 20.2c

FIGURE 20.2. Migration of a Positive Hole to the Surface of a Silver Bromide Grain (Movement from Right to Left)

overall positive charge to the lattice, but concentrated in the vicinity of the neutral bromine. Such a situation is known as a "positive hole." If negative charges migrate toward the positive hole, as shown in Figure 20.2 a, b and c, the positive hole moves in the opposite direction and eventually reaches the surface of the crystal. In the meantime, the released electron wanders through the crystal and eventually is trapped, or captured, by an impurity and is capable of attracting and reducing a wandering silver ion according to equation (5b). As a second photon interacts, the process is repeated. Another positive hole (bromine atom) is formed and moves to the surface. Ultimately, two positive holes get together to form a bromine molecule:

$$Br + Br \rightarrow Br_2 \qquad\qquad [7]$$

The presence of several silver atoms caused by this process is sufficient to sensitize the entire grain (lattice) to photographic development, and thus only a few (as few as four silver atoms out of 10^{10} silver atoms) constitute the **latent image**.

20.4. Photographic Development

Once the latent image is formed by exposure of the film, the next step in the process is development. Development may be defined as the process which transforms the invisible latent image into a visible image of elemental silver, the so-called **silver image**. You recall that equation (5b) indicated that the fundamental photographic process was the reduction of silver ions to silver atoms. Development accomplishes this by the interaction of a mild reducing agent with the silver ions. Reade accidentally discovered that gallic acid, with the formula

was such a suitable reducing agent and, interestingly enough, not much improvement on this basic formula has been made since. Modern organic developers are all some variation on the same theme as gallic acid, and are listed in Table 20.1 in order of increasing activity.

A typical development reaction is the following:

Hydroquinone, shown here on the extreme left, is a relatively mild developer. Pyrogallol is about sixteen times more vigorous, and metol and phenidone are about twenty times more vigorous than hydroquinone.

TABLE 20.2
SOME COMMON PHOTOGRAPHIC DEVELOPERS

Trade Name or Synonym	Formula	Chemical Name
Para-phenylenediamine	H_2N—⟨benzene ring⟩—NH_2	1,4-Diaminobenzene
Hydroquinone (Q)	HO—⟨benzene ring⟩—OH	1,4-Dihydroxybenzene
Pyrogallol Pyro Pyrogallic Acid	HO—⟨benzene ring⟩—OH, OH	1,2,3-Trihydroxybenzene
Metol (M) (as sulfate salt)	HO—⟨benzene ring⟩—N(H)—CH_3	N-Methyl-p-aminophenol
Phenidone	⟨pyrazolidone ring structure⟩	1-Phenyl-3-pyrazolidone

Although the developing process looks somewhat simple, the formulation of developers, as evidenced from their commercially available array, is not. In addition to the reducing agent itself, all developers need an accelerator, a preservative, a restrainer, and of course a solvent, usually water. An accelerator is necessary because most reducing agents react very slowly with silver halide emulsion, and the reaction could sometimes take hours if an accelerator were not present. Suitable accelerators are certain alkaline materials that are powerful enough to speed up the reduction process but not powerful enough to consume the gelatin substrate. Such chemicals are sodium carbonate (Na_2CO_3) and potassium carbonate (K_2CO_3), and on the less powerful side, sodium tetraborate, or borax ($Na_2B_4O_7$) and sodium metaborate ($NaBO_2$). Since carbon dioxide gas, CO_2, liberated from the reaction of the carbonates mentioned above tends to form blisters on the film, the latter two accelerators are preferred.

Since all photographic developers contain reducing agents, and all reducing agents have a tendency to become oxidized, even by atmospheric oxygen, it is necessary to add a preservative to the mixture which will prevent premature oxidation of the reducing agent. Such a preservative should have an affinity for atmospheric oxygen and become oxidized when in contact with it, so as to protect the developer, but it should not become oxidized when in contact with the film. Sodium sulfite, Na_2SO_3, is an ideal reagent for this purpose and is universally used as a photographic developer preservative. It reacts with atmospheric oxygen according to the following equation:

$$2Na_2SO_3 + O_2 \rightarrow 2Na_2SO_4 \qquad [9]$$

forming the water soluble, relatively inactive sodium sulfate, which can be washed away on completion of the development process.

Unfortunately, the presence of alkaline material in the developer causes fogging of the negative, and so the alkalinity must be lowered by adding a chemical that causes an acid condition in the solution. Halide ions, also formed in the development process itself (Equation [8]) are ideal for this, so all photographic developers also contain some halide salt, usually potassium bromide, in small amounts (about 1-2% by weight of the total dry ingredients). Table 20.2 lists the proportions of the various ingredients in some common photographic developers.

TABLE 20.2
RELATIVE AMOUNTS OF DRY INGREDIENTS IN SOME COMMON PHOTOGRAPHIC DEVELOPERS

Dry Ingredient (% by weight)	Agfa/Ansco 40 (General Use)	Agfa/Ansco 12 (Fine Grain)	Agfa/Ansco 22 (High Contrast)
Metol	3.7	5.7	0.8
Sodium sulfite	43.9	88.5	38.5
Hydroquinone	6.1		7.7
Sodium carbonate	43.9	4.0	48.2
Potassium bromide	2.4	1.8	4.8

Although all developers must contain some amount of reducing agent, preservative, alkaline salt and halide salt (commonly called a restrainer), by refining the types and the amounts of the ingredients, some special effects can be produced. For example, there are some photographic developers which, in the process of becoming reduced, produce enough restrainer that development of portions of exposed silver halides at the edge of the image is entirely prevented, thus yielding extremely sharp images. Other developers allow for some partial reduction of the unexposed silver halide, thus producing more subtle gradations in the image. Although grain size is fixed during manufacture, some developers can dissolve the silver image slightly, thus producing a fine-grain result in a fast film.

Developing time is also dependent upon the temperature. Most manufacturers provide a developing time chart with their film.

When the developing process is completed, it is necessary to stop further development by placing the film in a stop bath. Although plenty of ordinary water will do this job, some photographers prefer to use a dilute solution of acetic acid (3-4% w/v), since the acid will neutralize the alkaline conditions of the developer and thus inhibit further development.

20.5 The Latent Image Revisited

You recall that when the latent image was formed, only about four silver atoms were required per grain in order to sensitize the entire grain of silver halide. In the development process, each sensitized grain, regardless of size, is reduced to silver image by the developing agent. This is an all-or-nothing event. Each grain of silver halide is "autonomous," if you will, and will be reduced completely, if sensitized, regardless of how large it is. This fact can work both for and against the photographer, and the choice of a film with the proper grain size is important for achieving certain effects. For instance, if the film of choice contains fairly large grains of silver halide, then exposure by, say twenty, photons might activate about five grains of the halide. A film emulsion with a smaller grain size will have the same number of grains activated by the same number of photons, but since the grain size is smaller, a smaller area of film will have been utilized to form the latent image. Thus, a silver image less rich in actual silver will be formed upon development. The smaller grained film will require more light (more exposure time) at constant aperture setting in order to produce a latent image of the same intensity as the larger grained film. For this reason, the large grained films are sometimes called "fast films." However, although they require less exposure time, they produce images which are "grainy" and with much poorer image resolution (sharpness) than the images produced by slower film with smaller grain size.

We must keep in mind, also, that the processes which serve to sensitize the grains of silver halide in the film are all reversible. Given enough time, the positive holes and wandering electrons can recombine, thus yielding no latent image at all. This event takes place when film is exposed over a long period of time in very dim light, reducing the sensitivity of the emulsion under these conditions. One advantage of this fact is that the chances of fogging the film in the darkroom are almost nil if the light is sufficiently dim.

One might think if long exposure in dim light reduces the sensitivity of the film emulsion, then short exposure to very bright light should enhance it. Actually, this is not the case, again because of the reversibility of the equations governing the formation of the latent image. In the first instance, time was on the side of reversibility; in this instance, it is the population of positive holes and wandering electrons that increase the chances of reversibility. Short exposure to numerous photons releases many electrons at once, and since so many electrons are wandering around on the emulsion grain, the probability of recombination with positive holes is greatly increased. Thus, emulsion sensitivity is decreased by this condition as well. As with so many other cases in both chemistry and life, the pathway between the two extremes will produce the correct results.

20.6 Fixing

Once the silver image is formed, the emulsion is not yet completely processed. Some unexposed and, therefore, unreduced, silver halide still remains on the film and must be removed. Otherwise, further exposure to light will simply cause these unexposed portions to darken, thus destroying the dark-light contrast which comprises a photographic negative. Since the silver halide is insoluble in water, it cannot be removed by mere washing. Reagents must be added which render the silver halide soluble, and thus removable. The most common of these fixing agents are sodium thiosulfate (old name = sodium hyposulfate, from which the common name "hypo" is derived), $Na_2S_2O_3$, and sodium ammonium thiosulfate, $NaNH_4S_2O_3$. It is thought that the fixing reaction occurs in two steps by the formation of a silver-thiosulfate complex:

$$AgBr + Na_2S_2O_3 \rightarrow Na(AgS_2O_3) + NaBr \qquad [10a]$$

$$Na(AgS_2O_3) + Na_2S_2O_3 \rightarrow Na_3[Ag(S_2O_3)_2] \qquad [10b]$$

The final product is water soluble and can be washed away, leaving a completely processed negative behind. The fixation bath usually also contains a hardening agent such as potassium or chrome alum which prevents the soft gelatin coating from being damaged, an acidic buffer such as acetic acid or sodium acetate which prevents the fixing bath from becoming alkaline, and the preservative, sodium sulfite.

20.7 The Negative

Although you are used to seeing silver in a very smooth and highly polished lustrous form, the silver image that constitutes the photographic negative appears quite different. Here the silver has precipitated out in such fine grains, and is also so densely packed, that the surface is extremely rough, and thus multiple reflection and absorption of light takes place to such an extent that almost all the light striking the surface of the negative at the places where an image has been formed is absorbed. However, if you rotate the negative back and forth in ordinary light, you can also see the specular reflectance component of the white light. (Review the material in Chapter 9.2.)

If you have ever processed photographic film yourself, you know that one of the most important steps is the thorough washing and drying of the negative before printing. Remember that no chemical substance is stable indefinitely, and the chemicals that go into photographic processing are no exception. If even a tiny fraction of any of the processing chemicals is left behind on the negative, they will eventually decompose, react with the silver image, and cause deterioration of the negative. To prevent this deterioration, washing with chemically pure water is essential. Some drinking waters contain all kinds of chemicals which will interact with or dry out on the negative to cause spotting, and must be avoided. The washing procedure is logarithmic in the proportions of chemical removed. For example, if it takes one minute to remove 90% of the material, then two minutes of washing will remove 90% of the remaining 10%, and three minutes of washing will remove 90% of the remaining 1%, and so on. Even an infinitely long washing period can never remove all of the impurities; some will always remain to shorten the useful life of the negative. Furthermore, when washing prints, even more time is required because the chemicals are absorbed by the paper.

Drying of the negative is also important. It is common practice to first wipe off excess water with a squeegee since the water itself may contain chemicals which will spot the negative and show up in the print. Air drying of the negative is necessary to allow the gelatin, which contains up to ten times its weight in water, to release it to the atmosphere. If the gelatin has first undergone a hardening process, then it is possible to hasten the drying by using heated, forced air around the negative.

20.8 Reversal Processing

Sometimes, instead of making prints from the negative, it is desirable to make slides for projection. However, projection of the negative itself is undesirable because a true image is not seen. Projection of the negative can be remedied by playing a rather ingenious trick after removing the film from the acid stop bath. At this point, the unexposed silver halide, which **would constitute the positive image if it could be developed,** is still on the film. The trick is this: instead of placing the negative in the fix bath, it is placed instead in a bath which dissolves away the newly formed silver image - the negative portion of the film, if you will. This is accomplished with a solution of either sodium dichromate, $Na_2Cr_2O_7$, potassium dichromate, $K_2Cr_2O_7$, or potassium ferricyanide, $K_3Fe(CN)_6$. Once the silver image is gone, the remaining silver halide can be exposed by light fogging or by chemical means and subsequently developed and fixed in the normal manner. The result is a new negative which corresponds to the positive original image. It can be projected directly upon a screen.

Rapid processing by such a system as the Polaroid Land™ camera utilized a similar technique and was developed in 1939 by Edwin Land (1909-1991). In this process, the normal negative and a paper support backing which contains microscopic crystals of silver, or other metallic, sulfide, are in close contact with one another. After exposure of the negative, the "sandwich" is passed between special rollers which breaks a pod containing all of the ingredients needed for developing and fixing. This so-called mono-bath goes to work and develops the negative but, at the same time, the fixing agent dissolves off the unexposed silver halide. The whole assembly is held in a highly viscous plastic-like liquid which does not allow the $Na_3[Ag(S_2O_3)_2]$ complex formed according to equation [10b] to diffuse. Rather, it is received by the backing paper, and the minute crystals of metallic sulfide catalyze the precipitation of a silver image from a complex directly onto the backing paper. As soon as the process, which takes about one minute at room temperature, is complete, the amateur photographer strips off the negative and *voila!,* a direct positive print. Of course, this process is quite outdated now since most people who use "instant" cameras use automatic models which zip out pictures rapidly and where the film requires no stripping off of negatives. The images are also in color, a subject we will be dealing with shortly. (Personal Note: Dr. Land received more patents for his inventive work than any other American inventor except Thomas Edison. His now-obsolete, but once popular instant camera was inspired by his daughter's wish to be able to see a photograph the moment it was snapped. That wish is now more than fulfilled by the modern digital camera which requires no chemical processing.)

20.9 Printing

Since the chemistry of printing a positive is virtually identical to that of the formation of photographic negatives, we will not spend much time on it. Printing involves exposing a sheet of print paper through the negative and carrying out the same development processes mentioned before. However, there are several differences in the technique. First, the print paper can be divided into two varieties, the printing out paper which requires only exposure and no developing and is used for photographic proofs, and normal print paper which is exposed, and processed, in the usual fashion. The former type is undesirable for a permanent record because it darkens upon standing. The other difference is that whereas prints of the same size as the negative can be made by contact printing, prints of larger size can be made by increasing the distance between the negative and the print paper. This can be done in a controlled fashion by using an enlarger.

20.10 Supplemental Note

While much of the chemistry and the photographic processing involved in this chapter may seem to have become completely obsolete with the advent of the digital camera, there are still many photographic applications that require the good, old-fashioned method. Chief among them are the needs of professional photographers who need the infinite variations provided by different photographic films and concomitant developing agents to achieve their artistic results. There are also numerous medical and forensic applications that require film photography. So although it now occupies only a niche market and some leading corporations in the photography business such as Eastman Kodak have declared bankruptcy, traditional photographic processing will be with us for a long time to come.

In addition, the scientific principles that allow us to obtain and process black and white photographs apply also to color photography (with some refinements), and it is these principles that enable us to do four-color printing, presumably a need that will also remain with us for the duration.

20.11 Selected Readings

Baines, Harry *The Science of Photography,* 2nd Ed.; John Wiley and Sons: New York, 1967.

Bunting, R.K. *The Chemistry of Photography;* The Photoglass Press: Normal, IL, 1987.

Frizot, Michel. *A New History of Photography*; New York: Konemann: New York, 1998

James, T.H., Ed. *The Theory of the Photographic Process,* 4th Edition; The Macmillan Co.: New York, 1977.

Larmore, L. *Introduction to Photographic Principles;* Dover Books: New York, 1965.

Mitchell, Earl N. *Photographic Science*; John Wiley and Sons: New York, 1984.

Chapter 21

Color Photography

For mem'ry has painted this perfect day
With colors that never fade...

Carrie Jacobs Bond

CHAPTER 21
COLOR PHOTOGRAPHY

21.1. The Range of Sensitivity of Photographic Film.

Now that we have more or less disposed of the highlights of black and white photography, we come to a topic that is extremely important in understanding the differences in black and white film and color film. Although black and white photography may seem to you to be rather complicated, color photography is even more so. One of the most important factors involving color photography is simply making the film sensitive to the complete range of the visible spectrum.

Up to this point, we have not said anything about the spectral sensitivity of film. We have simply said that silver halides are light sensitive and let it go at that. However, silver halides are not naturally sensitive to all wavelengths of visible light, but only to the high energy, short wavelength end of the visible spectrum (the blue-violet region, you may recall) and to the ultraviolet region of the spectrum. As a result, an ordinary unsensitized emulsion does not respond to light in the red-yellow-green region of the spectrum, and objects reflecting these colors appear grayish in the finished black and white film.

It was Hermann Wilhelm Vogel (1834-1898), in 1873, who first discovered that a yellow dye could make photographic film more sensitive to the green region of the spectrum, but it was not until a generation later, in 1904, that the production of "panchromatic" film was made possible by Hoechst chemist Benno Homolka's (1860-1925) discovery of pinacyanol, a dye that was capable of extending the sensitivity of the silver halides to the entire range of the visible spectrum. Figure 21.1 depicts the light sensitivity curve of ordinary silver halide and the sensitivity that can be achieved for color photography by addition of blue-sensitive, green sensitive and red-sensitive dyes to the film emulsion. Since the range of maximum absorbance (between 300 and 400 nm for silver halides) indicates the range of maximum sensitivity of light, silver halides will tend to absorb (be sensitive to) light mainly in the 300-400 nm range. The blue, green and red-sensitive dyes have absorption maxima at longer wavelengths, and dyes have been developed which can extend the range of sensitivity of film to around 1,300 nm, well into the infrared region of the spectrum. Table 21.1 represents the structural formulas for some representative sensitizing dyes.

A general theory of silver halide sensitization involves the idea that the sensitizing dye acts as an intermediate between the photon and the silver halide grain in the formation of the latent image. Photons of low energy striking a film laced with sensitizing dyes do not have enough energy to knock an electron loose from the silver halide grain and create a positive hole. However, they do have enough energy to remove an electron from the sensitizing dye, thus effecting what ultraviolet light can do directly, *viz.*, create a mobile electron in the silver halide lattice. The dye is then capable of regaining its electron from a nearby halide ion, thus creating a positive hole. Now we are back in familiar territory, for the normal processes of latent image formation can take over from this point. However, a good deal more work is involved in devising a method for recording permanent images in color.

21.2. The Forerunners of Modern Color Photography

The first feeble step in producing an image in color occurred in 1777 when the Swedish chemist, Wilhelm Scheele (1742-1786, best known for his discovery of chlorine) discovered that the various

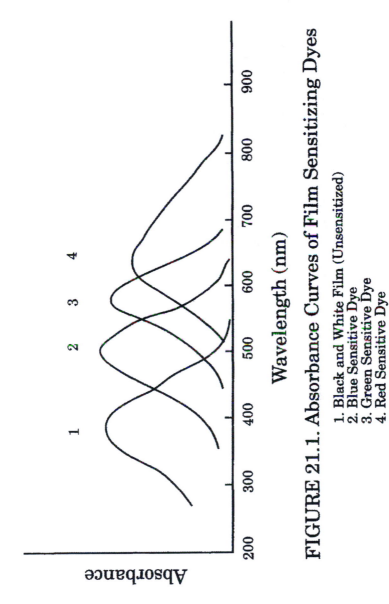

FIGURE 21.1. Absorbance Curves of Film Sensitizing Dyes

1. Black and White Film (Unsensitized)
2. Blue Sensitive Dye
3. Green Sensitive Dye
4. Red Sensitive Dye

TABLE 21.1
STRUCTURES OF SOME TYPICAL SENSITIZING DYES

Basic Structure

n	Color of Dye	Color Sensitivity	Name of Dye
0	Yellow	Blue	Cyanine
1	Magenta	Green	Carbocyanine
2	Cyan	Red	Dicarbocyanine

colors of visible light were not equally effective in darkening silver halides. He found that light on the blue-violet end of the spectrum was quite effective, that green light in an intermediate wavelength range was less effective, and that red light with long wavelengths hardly affected the silver halide at all.

The next step in the development of color photography came in the field of color-vision theory. The most important structure in the human eye for the perception of color is the retina, which contains the cone cells responsible for color vision. One of the earliest theories of color vision (1802) was that of Thomas Young (1773-1829) elaborated upon by Hermann von Helmholtz (1821-1894) around 1852. The combined Young-Helmholtz theory postulates that since the retina responds in at least three different ways to different colors, there must be three different kinds of receptors, each of which is sensitive to a particular portion of the visible spectrum. It took more than a century to obtain the cone-specific spectrophotometric information that showed that there are indeed three types of cones, each of which contains one of three light-sensitive pigments, corresponding to the additive primary colors, red, green and blue. According to the Young-Helmholtz theory, we perceive red because red light striking our eyes is capable of stimulating the red-sensitive cones; we perceive magenta because both blue and red cones are stimulated simultaneously; we perceive white when all the cones in a region of the retina are stimulated simultaneously; we perceive black when none of the cones is stimulated.

James Clerk Maxwell, who was also responsible for the color discs mentioned in Chapter 12, was the first to apply this new theory to the possibility of producing a colored photographic image. He reasoned that if an image were photographed three times through different colored filters, red, green and blue, and if the resulting photographs were developed and projected through filters of the same color and superimposed on one another in register on a screen, a full-color effect should be achieved. This would occur because in each case, only the red, green and blue components would be recorded in each of three images; projection in register would allow combinations of colors reflected from objects which contained more than one, all, or none of these primary colors. Maxwell's color photograph did indeed produce the expected results. His experiment was an application of the additive color primaries diagram shown in Figure 12.1. However, we are still a long way from modern color photography.

21.3. Color Photography

Although Maxwell's original experiment is still being practiced today by first analyzing the image by making three separate images, and then synthesizing the final colored image by bringing the separate color images in register, the process is both expensive and inconvenient. Its use is limited today to producing high quality color prints for publication. The most common form of color photography today is that done by the subtractive method.

Figure 11.6 depicts the relationship between the subtractive color primaries and their complements. Recall that when light passes through a colored filter, the light that is transmitted is the light that remains after the subtractive process has taken place. Thus, if white light is passed through a magenta-colored filter, magenta is perceived because its complement, green, has been subtracted from the white light. Similarly, a yellow filter will subtract out its complementary color, blue, and a cyan filter will subtract out its complementary color, red. However, the three subtractive primaries are actually color combinations. Cyan is a combination of green and blue, magenta of red and blue, and yellow of red and green. These combinations are very important in understanding how a colored photograph is produced by the subtractive process. These facts will be used in devising the film for color photography and the process for developing that film.

21.4. Color Photographic Film

Although it has been estimated that the human eye can distinguish between several million hues, each of these hues is merely a combination of minutely different proportions of the three primary colors, red, green and blue. If a color photographic film can be manufactured to record these subtle differences, but all reduced now to the marvelously simple three-color system, it should be able to record many of the nuances of color perceived by the eye.

The way this has been done is in the formulation of the color film tripack. In this film, not one, but three separate layers of silver halide have been coated onto a film base, but each layer has been made sensitive to only one color by the addition of the light-sensitive dyes mentioned previously. The red-sensitive layer is the first to be laid down on the film base, and contains not only ordinary silver halide, but also a dye that is sensitive to, *i.e.,* absorbs, red light. Remember that if a substance absorbs red light, it reflects or transmits the other colors, *viz.,* green and blue. Thus, the actual color of the red-sensitive dye that has been added to the emulsion must be cyan. Next, the green-sensitive layer containing silver halide and a green-sensitive dye of the type represented in Table 21.1 (what color is this dye?) is laid down, and finally, the blue-sensitive layer goes on top. Notice, now, that the top layer is sensitive to the most energetic range of visible light, blue. The middle layer is sensitive to the intermediate energies of visible light, and the bottom layer is sensitive to the least energetic range of visible light, red. This arrangement is not accidental. When light of various colors strikes the tripack film, blue light will be absorbed at the surface, and will not go on to expose the other layers which are also sensitive to blue light. However, just in case some blue light does get through the top layer, a yellow filter, which absorbs blue, is placed between the blue-sensitive and green-sensitive layers of the film. Thus only green and red light can travel past the yellow filter. The green light will be absorbed in and expose the green-sensitive layer. Even if some of it passes through into the red-sensitive layer, it does not matter because neither the silver halide nor the red-sensitive dye in this region is sensitive to the green light and it will remain unabsorbed. Finally, the red light will be absorbed in and will expose the red-sensitive layer. Remember that if only magenta light strikes the film, then only the layers corresponding to absorption of magenta light, that is, the blue and red-sensitive layers will be exposed. The green layer will be unaffected by magenta light. In most color films that are exposed through the front, a film backing is also present under the red-sensitive layer that absorbs all the unabsorbed photons to prevent reflection back into the tripack. If a ray of light is reflected, it will expose the film twice in two different places according to the laws of reflection, and the image will appear fuzzy.

21.5. Color Processing

Color and black and white film processing differ in at least three respects:

1. . Black and white processing had the silver image as its final product; in color processing, the silver image is unimportant. The **dye image** is the important product.
2. In black and white processing, the reduction product, silver, was the important product; in color processing, it is the oxidation product, a dye formed between the color developer and certain chemicals called "couplers," that is important.
3. Black and white developers were rather simple mixtures, and processing could be carried out in one step; in color processing, there are multiple steps, and almost as many different developers.

First, let us look at these new compounds, couplers, and see what relationship they have to color film processing. These couplers were discovered in 1912 by Rudolf Fischer (1881-1957), and as their name indicates, they couple with other substances to form larger molecules. Their importance lies in the fact that each molecule of the coupler is capable of coupling with one molecule of color photographic developer to form a dye molecule. The dye molecules formed comprise one of the components of the dye image in the color photographic print or negative. In the process of coupling, the photographic developer reduces the silver halide of the latent image to metallic silver and becomes oxidized as a result. Thus, this entire class of reactions is known as oxidative coupling. The reaction of some typical couplers with the color photographic developer, N,N-dimethyl-p-phenylenediamine, are given as follows: Varying proportions of the three dyes formed in equations 21.1, 21.2 and 21.3 embedded on the color negative or color print, make up the dye image we see. How this is done depends upon the particular process used. The tripack color film (also invented by Fischer, and now trademarked by several film corporations) incorporate the couplers right in the film emulsion. The coupler of equation 21.1, α-naphthol, or one of its chemical analogs, is incorporated into the red-sensitive layer. The coupler in equation 21.2, benzylacetanilide, or equivalent, is incorporated into the blue-sensitive layer, and the coupler of equation 21.3, cyanoacetylcoumarone, or equivalent, is incorporated into the green-sensitive layer. Of course, it takes a little chemical know-how to prevent these couplers from wandering from layer to layer. This can be accomplished by attaching long chain hydrocarbon groups to them to act as "ballast" groups.

Let us now imagine that we have exposed our color tripack to magenta light. As we mentioned earlier, only the green-sensitive layer will remain unexposed to magenta light. Next, we add developer (N,N-dimethyl-p-phenylenediamine, or similar reagent), and reaction 21.1 takes place in the red-sensitive layer to produce a cyan dye in the layer, and reaction 21.2 takes place in the blue-sensitive layer to produce a yellow dye. Of course, silver image as well as dye image is also formed, so it must be bleached away by

[21.1]

Developer + Coupler (α-Naphthol) + 4Ag⁺ → Cyan Dye Image + 4Ag + 4H⁺ Silver Image

[21.2]

Developer + Coupler (Benzylacetanilide) + 4Ag⁺ → Yellow Dye Image + 4Ag + 4H⁺ Silver Image

[21.3]

Developer + Coupler (Cyanoacetylcoumarone) + 4Ag⁺ → Magenta Dye Image + 4Ag + 4H⁺ Silver Image a wash solution of potassium dichromate or potassium ferricyanide. Now, we must place the film in a fixing bath to dissolve off the unexposed silver halide (largely in the green-sensitive layer), and after washing and drying, our film is ready for viewing. Since a yellow dye was formed in the blue-sensitive layer, and a cyan dye was formed in the red-sensitive layer, when these images are viewed in register in transmitted light, the image we see will be green. Since our original image was magenta, this means we have succeeded in producing a negative in the complementary color of the original. Producing this color negative will involve exposing color print paper through our green negative and, after processing, the image will appear in the color complementary to green, namely, magenta. Thus we will have recorded the magenta image.

Direct positives can also be produced in color by reversal processing, but this involves using film which contains no couplers, and introducing the couplers in the development process. After exposure of reversal-type film to a magenta image, black and white development takes place in the usual way, leaving silver image largely in the red- and blue-sensitive layers, and virtually no reduced silver in the green-sensitive layer. Next, three color developments take place. First, the red-sensitive layer is exposed to red light trough the back of the film (thus making sure that neither of the other two layers is exposed). Then α-naphthol and N,N-dimethyl-p-phenylenediamine are added and oxidative coupling takes place in the red-sensitive layer to produce a cyan dye and metallic silver. If our magenta object were pure magenta, then little or no cyan dye will form because the silver halide will have already been exposed to the red component of the magenta image, and reduced to silver in the preceding black and white process. Next, the film is exposed through the front to blue light, and color developer plus benzylacetanilide, or equivalent, is added. Reaction 21.2

takes place in the blue-sensitive layer to produce yellow dye but, again, very little of this will be produced if the original object was mainly magenta. Finally the green-sensitive layer is exposed and developed with the color developer and cyanoacetylcoumarone, or equivalent, and a magenta dye forms in the green-sensitive layer. Since this layer was largely unexposed originally, a good deal of magenta dye will form and our original magenta object will appear directly on the transparency. This entire scheme is outlined in both of Thirtle's papers given in the reading list at the end of this chapter.

We should keep in mind that the colors we see in color prints, transparencies and slides are the colors of the dyes formed between the reaction of the couplers with the color developer. Thus, the color can only approximate the "true" color of the original object. A great deal of research has gone into developing couplers that will render colors as we see them. It should also be obvious that since the construction and processing of color film is so much more involved than the analogous processes in black and white photography, color photography is necessarily more expensive.

21.6. A Word About Digital Photography

Earlier in our introduction to photography, we mentioned that it was the convergence of physics and chemistry that allowed for the development of the permanent recording of images by the use of light. With the advent of digital photography, the discipline swung back into the bailiwick of the physicists, since digital photography is almost purely based on physics, There are many fine tutorials on this subject on the internet, including not only the principles, but also the techniques of obtaining fine digital images.

Since this area lies cearly outside the purpose of this book, we merely enumerate some good websites to follow up on if you are interested in pursuing this topic.

http://www.macro-photography-for-all.com/index.html

http://www.macro-photography-for-all.com/digital-color-photography.html

http://www.cambridgeincolour.com/

21.7. Selected Readings

Falk, David, *et al. Seeing the Light;* Harper and Row: New York, 1986.

Kapecki, Jon; Rodgers, James. Color Photography. In *Kirk-Othmer Encyclopedia of Chemical Technology,* Vol. 19, 2007, 231-272

Keller, E. Images in Color. *Chemistry* **1970**, *43* (Dec.), 6-10.

Thirtle, J.R. Inside Color Photography. *CHEMTECH* **1979**, *9*, 25-35.

Thirtle, J.R.; Zwick, D.M. Color Photography. In *Kirk-Othmer Encyclopedia of Chemical Technology,* Vol. 5, 1964, 812-845.

Chapter 22

Ceramics, Glasses and Glazes

Hath not the potter power over the clay?

Romans 9:21

CHAPTER 22
CERAMICS, GLASSES AND GLAZES

Ceramic materials constitute some of the earliest artifacts of humankind. This is not surprising since ceramics are made of the stuff of the earth, the clay-like material that can be found almost anywhere. It is even less surprising when we consider that all ceramic materials are noted for their durability, hardness and resistance to attack from heat and corrosive materials.

The practice of working ceramics started out as a practical one oriented toward providing utilitarian objects such as bowls, lamps and storage vessels. It quickly became an art form, as one can see by strolling through almost any museum and viewing the highly decorated ceramic materials of ancient civilizations that survive to this day. In our own time, ceramics has also become a major technology, and ceramic materials are now used in the production of heavy clay products, industrial abrasives, porcelain enamels, electronic materials, semiconductors, and a host of other objects and devices needed by our society. In this chapter, we will orient our discussion of the nature of ceramic materials to their use as artists' materials. Since glasses and glazes are so closely related to ceramics, they will also be included in this chapter.

22.1. The Origin and Composition of Clay

Any discussion of ceramic material must begin with a consideration of their basic raw material, clay. Clay was formed by the action of wind and water upon the earth's crust long before human beings evolved on this planet. This natural clay-forming process continues today by the slow weathering of minerals originally formed from among the most abundant elements in the earth's crust.

Most of the earth's crust is composed of a relatively small number of elements. Oxygen, silicon and aluminum represent 82.7% by weight of the crust, and the ten most abundant elements (see Table 22.1) make-up over 99% of its weight. The remaining 80 naturally occurring elements constitute only 0.8% of our world.

Oxygen and silicon, the two most abundant elements, tend to form a group of compounds called silicate minerals. Naturally, since silicon and oxygen are so abundant, one would expect their compounds to be the most abundant also, as indeed they are. Silicates are not simple materials. They contain a basic unit in which each silicon atom is bonded to four oxygen atoms in a tetrahedral arrangement (Figure 22.1). One such isolated unit would represent the orthosilicate ion, SiO_4^{4-}. Examples of minerals containing the orthosilicate ion are zircon, $ZrSiO_4$, and forsterite, Mg_2SiO_4.

More complex silicate structures generally are found in which the tetrahedral silicate units share oxygen atoms. Thus, two tetrahedral units with a shared oxygen would have the formula, $Si_2O_7^{6-}$. Compounds containing this unit are called disilicates. It is possible to have single-stranded silicates in which the neighboring tetrahedra each share two oxygen atoms to form a long single chain or a long double chain. Both types of such chains are called asbestos. In a similar manner, it is possible for each tetrahedral unit to share three, or even all four, oxygen atoms, and form two-dimensional sheetlike structures, the micas, or three-dimensional arrays, such as quartz. These various possibilities are summarized in Table 22.2.

When silicates are formed, it is important to remember our chemical bookkeeping in writing their formulas. Silicon, in Group IV of the Periodic Table, contributes four valence electrons, and

TABLE 22.1
THE TEN MOST ABUNDANT ELEMENTS IN THE EARTH'S CRUST

Element	Weight %
Oxygen	49.5
Silicon	25.7
Aluminum	7.5
Iron	4.7
Calcium	3.4
Sodium	2.6
Potassium	2.4
Magnesium	1.9
Hydrogen	0.9
Titanium	0.6

FIGURE 22.1
REPRESENTATIONS OF THE ORTHOSILICATE TETRAHEDRAL UNIT

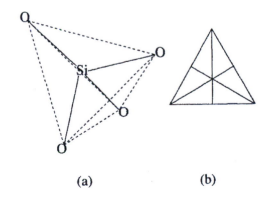

(a)　　　　　　(b)

TABLE 22.2 SOME REPRESENTATIVE SILICATE MINERALS

Structure	Formula	Type of Silicate Description	Examples
	SiO_4^{4-}	Discrete tetrahedra	Fe_2SiO_4 (Olivine) $ZrSiO_4$ (Zircon)
	$Si_2O_7^{6-}$	One oxygen atom shared between two tetrahedra	$Sc_2Si_2O_7$ (Thorveitite)
	$(SiO_3^{2-})_n$	Pyroxene (Asbestos Mineral)	$Na_2Al_2Si_4O_{12} \cdot 6H_2O$ (Levyne)
	$(Si_4O_{11}^{6-})_n$	Amphibole (Asbestos Mineral)	$Ca_2Mg_5(Si_4O_{11})_2(OH)_2$ (Tremolite)
	SiO_2	Four Si-O-Si bonds for each tetrahedron (tetrahedron represented with dashed lines is below the plane of the other four)	Cristobalite Quartz Tridymite

may be said to have a nominal oxidation number of +4. Oxygen, in Group VI, has two vacancies in its valence shell and tends to gain two electrons. Hence, the simplest formula for the silicon-oxygen compound is SiO_2, in which silicon's four valence electrons have been distributed between two oxygen atoms. Of course, this bookkeeping is somewhat fictional because the electrons are actually shared. However, this viewpoint is useful for the discussion that follows.

When silicon and oxygen form tetrahedra in which four oxygen atoms are covalently bonded to one silicon atom, each oxygen atom can only share one of silicon's four valence electrons, leaving each oxygen hungry for one more electron. Thus, the total electron-poor arrangement must look outside of itself for another source of electrons. Since metals tend to give up electrons quite readily, they are a handy supply, and so silicate tetrahedra are always found bound up with metal ions. Zircon, our earlier example of a simple orthosilicate, consists of one zirconium atom which has lost four electrons to four electron-hungry oxygen atoms in the silicate tetrahedron. Hence, in our bookkeeping scheme, we can write Zr^{4+} to indicate zirconium's nominal oxidation number, and SiO_4^{4-} to indicate the oxidation number of orthosilicate. Similarly, the $Si_2O_7^{6-}$ grouping has six negative charges (and metal ions with a total of six positive charges hovering in the vicinity), $Si_3O_{10}^{8-}$ has eight negative charges, and so forth. Silicate minerals, therefore, may be thought of as covalently bonded arrays of silicate tetrahedra interspersed with metal ions (cations) to keep the charge balance.

Since aluminum is the third most abundant element in the earth's crust (Table 22.1), we might expect that the initial formation of the silicate minerals might have been processed with the occasional inclusion of an aluminum atom in the silicate chain in place of silicon. This has indeed happened, and with tremendous consequence. Aluminum, in Group III of the Periodic Table, has only three valence electrons compared with silicon's four. This means that each time an aluminum ion, Al^{3+}, became incorporated into the silicate structure by displacing or taking the place of a Si^{4+}, the difference in positive charge had to be compensated for by the inclusion of additional cations such as Na^+, K^+, Ca^{2+}, Mg^{2+} and Fe^{3+} in the structure. The resulting minerals are called aluminosilicates, and among them are the feldspars with a typical formula of $K_2Al_2Si_6O_{16}$. The feldspars are the most abundant of all minerals.

Numerous explanations for the formation of the clay mineral, kaolinite, from feldspars have been advanced. It is believed that long exposure to atmospheric water and carbon dioxide may have resulted in the following typical reaction:

$$K_2Al_2Si_6O_{16} + CO_2 + 2H_2O \rightarrow Al_2Si_2O_3(OH)_4 + 4SiO_2 + K_2CO_3 \quad [22.1]$$
Feldspar   Kaolinite   Silica

This explanation is rather difficult to accept because of the drastic structural change necessary to form kaolinite from feldspar, a change that would probably need far more impetus than that available from atmospheric vapors at ambient temperatures. However, it is useful as a way of introducing the formula for kaolinite, the ideal clay mineral.

22.2. The Properties of Clay

However it is produced, pure kaolinite is a white mineral which occurs as small, thin platelets with an irregular hexagonal shape. It occurs in nature mixed with silica, SiO_2, feldspar, and a host of metallic oxides and organic materials which cause the clay to vary in color from yellow to red to black.

The structure of kaolinite accounts in large measure for the properties of wet clay as it is being worked, and for the properties of the ceramic material after it has been fired. In kaolinite, a silicate sheet is connected through ionic bonds to an aluminate layer containing Al^{3+} and OH^- ions, with water molecules (sometimes called lattice water) sandwiched between the two sheets as illustrated in Figure 22.2. When wet, these sheets cling to each other like wet playing cards, and with water as a lubricant, the crystals readily slide over one another. This accounts for clay's plasticity. Other factors such as temperature and the types of cations adsorbed onto the clay's surface also change its manipulability when wet. Still another, and relatively uncontrollable factor, is the aging process which improves clay's plasticity and workability. It appears that certain bacteria contained in the clay cause it to "ripen" over a period of several weeks with marked improvement in its properties. Aging is favored by keeping the clay at a relatively warm temperature and feeding the bacteria by adding a small amount of starch to the clay mass. When a new batch is being mixed, it is common to add a portion of previously aged clay as a "starter" in much the same way that sourdough bread starter is added to a new batch of bread dough.

22.3. The Firing of Clay

After the clay has been molded into the desired object, usually slightly larger than the projected final size, it is allowed to air dry to remove excess water. This is known as the "green" stage. At this stage, the kaolin platelets are like a dry deck of cards. The structure is extremely fragile and very porous. After drying, the

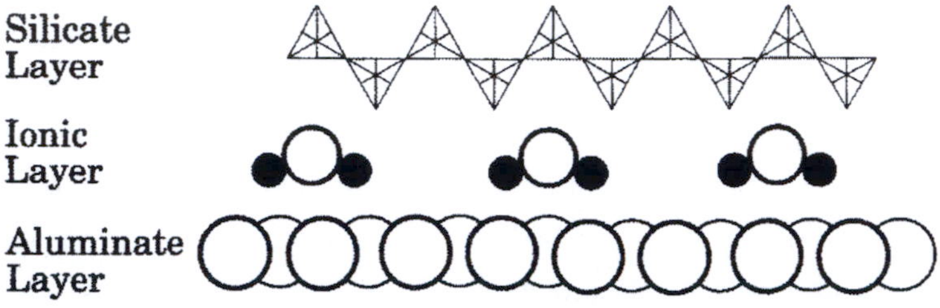

Silicate
Layer

Ionic
Layer

Aluminate
Layer

FIGURE 22.2. Structure of Kaolinite
(●◯● = Water)

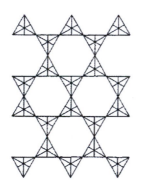

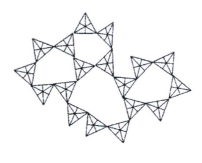

FIGURE 22.3a. The Long-
Range Order of a Crystal

FIGURE 22.3b. The Short-
Range Order of a Glass

△ = Silicate Tetrahedron

clay object is fired, sometimes at temperatures of up to 1450 °C. In the 100-125 °C range, residual water is removed. Between 350-525 °C, the lattice water is driven off, but without shrinkage of the object. Quartz, a normal component of clays, undergoes a change to a different crystalline structure at 573 °C. The organic matter which is also a component of most clays burns off, if a sufficient supply of air is present, between 200-1000 °C. Magnesium carbonate and calcium carbonate decompose to their respective oxides and carbon dioxide at 790 and 880 °C respectively. The calcium and magnesium sulfates decompose in the 1100 to 1300 °C range.

When a kaolin sample is fired to almost the complete dehydration point, a slight change in the structure results to give a new product, metakaolin. Further heating leads to a defect spinel-type structure at about 950 °C. In the 1000 to 1250 °C range, long, needlelike crystals of mullite, $Al_6Si_2O_{13}$, are formed, and these enhance the strength of the structure. Another quartz transition takes place at 870 °C, followed by still another at 1470 °C to form cristobalite. All of these solid phase transitions, which are summarized in Table 22.3, take place very slowly, so the firing process must also take place slowly to permit these changes to take place without damage to the clay body. During firing, the formation of mullite and glass from the feldspar and silica present serve to seal the entire structure together in a hard, strong mass. The degree of this glass formation, called vitrification, influences the porosity and strength of the final object.

Considerable shrinkage, between 10 and 15%, occurs during the firing process. This can be reduced by adding pieces of finely ground, already fired clay to the clay body before molding. The product of the first firing is called "biscuit."

TABLE 22.3

FIRING OF CLAY

Temperature (°C)	Event
100-125	Residual (unbound) water is removed; kaolin is transformed into metakaolin
200-1000	Organic material burns off
350-525	Lattice water (water that is part of the crystalline structure) is removed; no shrinkage occurs
573	Low quartz is transformed into high quartz
790	Magnesium carbonate, $MgCO_3$, decomposes to magnesium oxide, MgO, and carbon dioxide, CO_2
870	Quartz is transformed to tridymite
880	Calcium carbonate, $CaCO_3$, decomposes to calcium oxide, CaO, and carbon dioxide, CO_2
950	Metakaolin is transformed into a defect, spinel-type structure
1000-1250	Mullite, $Al_6Si_2O_{13}$, is formed from the feldspar present
1100-1300	Calcium sulfate, $CaSO_4$, and magnesium sulfate, $MgSO_4$, decompose
1470	Tridymite is transformed into cristobalite

22.4. The Properties of Fired Clay

At this point, we now have enough knowledge about ceramic materials to be able to define them. We usually know what ceramics are by instinct, and any person on the street would be able to give an example of a ceramic piece without necessarily being able to define "ceramic." Basically, we can define a ceramic as a combination of one or more metals or semimetals, such as silicon, with oxygen, linked by ionic and covalent chemical bonds into a polymer-like matrix. The five main types of ceramic pottery produced throughout the world at the present time are earthenware, stoneware, bone china, porcelain and vitrified pottery.

Since ceramic materials are rich in oxygen, they have no tendency to combine with more oxygen or any other element. Consequently, they are highly heat resistant, impervious to attack by nearly all chemicals and only very slightly soluble in water and other solvents. The nature and strength of the metal-oxygen bonds account for these properties.

However, since the fired ceramic article still consists essentially of kaolin platelets, it is a somewhat porous object. For this reason, it is necessary to coat fired pottery with a glaze which has three functions: 1) it makes the pottery nonporous and watertight; 2) it enhances the strength of the object; 3) it adds to the beauty of the object.

22.5. The Nature of Glasses

A ceramic glaze is a thin glass layer applied to a pottery surface and then fired in a kiln. Since the chemistry involved in glaze formation differs only slightly from that used in the manufacture of glass, it would be useful to examine the nature of glasses first.

What is the difference between a ceramic material and a glass? Initially, one can outline their differences by looking at their properties. Ceramics are shaped at room temperature, while glasses are shaped at elevated temperatures. Ceramics harden with application of heat; glasses harden upon cooling. Ceramics are composed of silicates and aluminates mixed with silica and some salts in minor quantities; glasses are largely silica mixed with agents which lower the melting point of the silica (fluxing agents). Ceramics are almost visibly porous; glasses are nonporous. Both are nonconductors of heat and electricity. Ceramics are micro-crystalline in structure; glasses are noncrystalline in structure.

This last-named property, the structural property, is really the heart of the matter. All silica is not necessarily glass, but only the silica that is non-crystalline in structure. The essential nature of glass is that it consists of relatively disordered silicate tetrahedra. Quartz, a non-glassy silicate, consists of repeating subunits of SiO_4^{4-} tetrahedra arranged in long-range order (Figure 22.3a) whereas a glass is an array of silicate tetrahedra characterized by a shorter range order (Figure 22.3b). Because of its great order, it is necessary to break many bonds at once in order to transform quartz into a liquid. Hence, it has one of the highest melting points known. However, because of its short-range order, glass melts much more easily. Fewer bonds need to be broken and the disorderly array is less thermodynamically stable to begin with.

22.6. The Properties of Glasses

Pure silica has a melting point of around 1700 $^{\circ}$C, a rather unreasonable temperature, and as a liquid, it is too stiff and viscous to work with even at any reasonable temperature. Thus, both melting point and viscosity must be reduced by adding fluxing agents such as metal carbonates to the material. Addition of sodium and calcium fluxes to silica gives rise to the so-called soda-lime glasses which are used extensively for window glass and ordinary glass objects. More specialized glasses can be produced by adding boron, giving rise to the borosilicate glasses, the most famous example of which is "Pyrex"^{TM} . Boron oxide is capable of lowering the melting point and the viscosity of the silica, but does not increase its tendency to expand on heating, thus making such glass "ovenproof."

The fluxing agents are able to lower the melting points of crystalline materials because they disturb the long-range order of the crystals. Crystals are held together by four types of bonds,

(1) metal-metal bonds,

(2) ionic bonds,

(3) van der Waals and hydrogen bonds, and

(4) covalent bonds.

The great strength and high melting points of ceramic crystals are due to the fact that they are held together mainly by ionic and covalent bonds. These are the bonds most difficult to break and hence require more energy in the form of heat in order to do so. When the long-range order of a crystalline material has been disrupted by heating so that molecules can slip and slide over one another, melting has occurred. Fluxes are foreign materials which cannot become part of the crystalline structure; hence they can only be bound to the portions of crystalline material surrounding them by very weak van der Waals and hydrogen bonds. Since these bonds are much weaker than the ionic and covalent bonds in the crystal, it takes less energy to cause melting. In general, impure crystalline materials have lower melting points than their purer counterparts.

22.7. The Desired Properties of Glazes

A glaze is formulated in order to impart beauty to an object and also to increase its overall strength and chemical resistance. Therefore, it must be fluid enough to fill the external pores of the ceramic piece upon firing, but viscous enough not to run off the piece in the process. In addition, a glaze is often used to hide the clay body from view. Therefore, ceramic glazes must contain fluxing agents for lower melting points, stiffening agents for increased viscosity, opacifiers for hiding the clay body, and coloring agents for added beauty. The most common fluxing agent is boric oxide, usually added as the mineral colemanite, since this form of boron-containing material is insoluble in water. Another advantage of using boron as a flux is that it also reduces the coefficient of expansion of the glaze so that on cooling, it will be in a state of compression. If the glaze is stretched as it cools, cracks, or "crazing," will develop on the piece. Aluminum oxide, alumina, is often used in small amounts as a stiffening agent. It not only keeps the glaze from running while in the molten state, but also reduces the tendency of the glaze to crystallize on cooling. Alumina is usually added in the form of fairly pure kaolinite. Tin(IV) oxide, SnO_2, is the most widely used opacifier.

A fluxing agent for glazes that has superior properties is lead oxide, PbO, in the form of litharge. Lead-containing glazes are usually smooth and clear, adhere well to the pottery surface and interact well with oxides used to provide color. While litharge was widely used for this purpose in the past, recent consciousness regarding toxic materials in the environment has led to its disuse.

A glaze is usually applied to pottery at the biscuit stage, although it may sometimes be applied at the green stage. The glaze itself is a carefully formulated mixture of silica, fluxing agents, stiffening agents and colorants suspended as finely divided particles in water. It can be described as a creamy, opaque slurry, or "slip," which may be applied by spraying or dipping the pottery. It is usually fired in the range 1060-1250 °C. The properly glazed piece of pottery has a smooth, glassy surface.

22.8. Colorants in Glazes

Coloring agents added to the glass melt in low concentrations include colloidal copper which produces a red ruby glass, iron polysulfide, which produces the familiar "beer bottle glass" color, chromium oxide, cobalt oxide and cadmium sulfide, which produce green, blue and orange glasses respectively. These same agents may be used in ceramic glazes, but the firing conditions may markedly affect the chemistry, and hence, the color, of the coloring agent.

Since the ceramic firing process often demands temperatures as high as 1400 °C, a glaze colorant must be able to stand up to these extreme temperatures. Early potters discovered that the compounds of copper and the iron earths were ideally suited for this purpose and used the former to produce turquoise blue and green colors, and the latter to produce yellow, green and brown. Progress in the development of high temperature colorants was slow. The iron earths and copper compounds remained the mainstay of these colorants until about the year 1200. Around this time, Chinese potters introduced red copper(I) oxide, Cu_2O, and later, yellow lead antimonate, $Pb_3(SbO_4)_2$, blue cobalt silicate, Co_2SiO_4, and manganese silicate, Mn_2SiO_4, which produces a purple brown color. The Moors introduced tin(IV) oxide, SnO_2, as a white colorant, and colloidal gold was known in Europe as a pink colorant from the latter part of the sixteenth century. An interesting colorant known from ancient times and still in use today as a high-grade glaze colorant of unsurpassed clarity and brightness is "Purple of Cassius." It is produced by adding a solution of tin(II) chloride, $SnCl_2$, to a very dilute solution of gold chloride, producing a precipitate of hydrated tin(IV) oxide, SnO_2, interspersed with finely divided elemental gold:

$$6H_2O + 3SnCl_2 + 2AuCl3 \rightarrow 2Au(s) + 3SnO_2 + 12HCl \quad [22.2]$$

This is an oxidation-reduction reaction in which the tin is oxidized and the gold is reduced according to the following half-equations:

$$Sn^{2+} + 2H_2O \rightarrow SnO_2 + 4H^+ + 2e^- \quad [22.3a]$$

$$Au^{3+} + 3e^- \rightarrow Au(s) \quad [22.3b]$$

The reduction product, solid gold, Au(s), in a very finely divided state, colors the resulting precipitate brown, purple or red, depending upon the original concentration of the solution.

As chemistry replaced alchemy in the mid-eighteenth century, progress in glaze colorants became more rapid. Chromium salts were introduced in France, and their use quickly spread to England and the rest of Europe. As more new materials became available, pottery manufacturers began to produce their own colorants and colorant formulations, which often took the form of "secret formula" books. Over the following century, glaze formulation gradually became a manufacturing specialty in its own right, and by the end of World War II, in-house manufacture of colorants by pottery makers had virtually ceased.

Although the amateur potter can produce the complete spectrum of glaze colors by the use of oxides or carbonates of only eight metals, chromium, cobalt, copper, iron, manganese, nickel, titanium and vanadium, numerous patents have been issued involving other elements such as zirconium, antimony, cadmium, gold, selenium and uranium. The formation of a glaze colorant in the firing process is not simple, and the study of phase diagrams at high temperatures and under various firing conditions has resulted in postulated mechanisms of color formation.

Brown, red-brown, and chocolate brown glazes can be formed by varying mixtures of the oxides of zirconium, iron, chromium and aluminum. Black glazes can be produced by varying compositions of the oxides of copper, manganese, iron, chromium, nickel and cobalt. Copper can produce blue or blue-green glazes if it is present in the glaze crystal lattice as copper(II), a state produced by firing in an oxidizing, or oxygen-rich, atmosphere. However, in a limited supply of air, copper(II) oxide may be reduced to red copper(I) oxide, Cu_2O, according to the following equation:

$$2CuO \rightarrow Cu_2O + 1/2\ O_2 \qquad\qquad [22.4]$$

Red Cu_2O may also be produced under oxidation firing conditions if a reducing agent such as silicon carbide, SiC, is included in the glaze:

$$8CuO + SiC \rightarrow 4Cu_2O + SiO_2 + CO_2 \qquad\qquad [22.5]$$

A beautiful blue color may also be produced by heating a mixture of zircon, $ZrSiO_4$, and vanadium pentoxide, V_2O_5. The production of this color has been the subject of much theoretical speculation, but it is generally agreed that incorporation of vanadium into the zircon lattice is responsible. Since the zirconium in zircon has an oxidation number of 4+, the vanadium, which in vanadium pentoxide has an oxidation number of 5+, must be reduced to V^{4+}:

$$V_2O_5 \longrightarrow V_2O_4 + 1/2\,O_2 \qquad\qquad [22.6]$$

This reaction takes place at around 700 $^\circ$C.

It has long been known that pink glazes could be produced by strong heating of the oxides of chromium and tin with fluorspar, CaF_2. It has been postulated that this happens by incorporation of chromium ions in the lattice of $CaSnSiO_5$, a compound called "tin sphene" which forms when SnO_2, CaO and SiO_2 are fluxed together. The well-known chrome-alumina pink is formed when chromium becomes incorporated into the lattice of alumina, Al_2O_3. Naturally occurring crystals of alumina with incorporated Cr^{3+} ions are called "ruby." Yellow glazes may be formed by mixtures of vanadium pentoxide with SnO_2 or ZrO_2. Orange glazes of varying hues can be made by incorporation of chromium and antimony ions in the rutile lattice of TiO_2. An increase in the chromium content gives rise to a red-orange, and an increase in antimony content yields an orange on the yellow side.

This brief discussion of glaze chemistry makes it all too clear that glaze formation is one of the more complex types of high-temperature chemical reactions. Many different reactions, in different phases, and under different conditions of firing, may take place. However, it is essential that the glaze be compounded in such a way that it fulfills exactly the required ultimate melted analysis, *i.e.*, with no starting material left over and unreacted. Such compositions, while the results of trial and error before the advent of chemistry, may now be precalculated on the basis of the series of chemical reactions which take place in the firing process. This is one reason why studies of the exact mechanism of glaze formation are so important.

22.9. Selected Readings

Bell, B.T. The Development of Ceramics. In *Review of Progress in Coloration and Related Topics,* Vol. 9, 1978.

Charles, R.T. The Nature of Glasses. *Scientific American* **1967**, *217* (Sept.), 126.

Denio, A. Chemistry for Potters. *Journal of Chemical Education* **1980**, *57*, 272-5..

Gilman, J. The Nature of Ceramics. *Scientific American* **1967**, *217* (Sept.), 112.

Orenstein, L. The Chemistry of Ceramics. *The Science Teacher* **1989**, *56*(April), 49-51.

Rasmussen, S. *How Glass Changed the World.* Springer: Heidelberg, 2012.

Chapter 23

Art Hazards

I have observed that nearly
all the painters I know...are sickly;
and if one reads the lives of
painters, it will be seen that
they are by no means long-lived....

Bernardino Ramazzini
De Morbis Artificum Diatriba, 1713

CHAPTER 23
ART HAZARDS

23.1. Safety in the Studio

Given the many accidents concerning the use of chemicals that have lately been highlighted in the news, safety has become an urgent concern in industry, in academic settings such as chemistry laboratories, as well as in artists' studios. Though the effects of working with toxic art materials was documented by Ramazzini as early as 1713, the first National Conference on Arts and Crafts Hazards was not held until October, 1978. Fortunately, the proceedings of this conference led to a rash of information in the form of publications (listed at the end of this chapter), and technical data sheets on materials and processes.

Reliable information should be accompanied by responsible practice. Listed below are the ten most important safety rules recommended by the Arts Hazards Project for artists and craftspeople:

1. Choose the studio location carefully. It should be located outside of the home, if possible, but most certainly away from the kitchen area and from children.

2. Use the least toxic materials and processes possible. It is possible to substitute less toxic solvents for those which are known to cause harm in many instances; certain techniques, such as wet-grinding of pigments, are less hazardous than others, such as dry-grinding;

3. Read labels carefully. Only products that are properly labeled, that is, that contain information regarding ingredients, directions for use, cautions to be followed and manufacturer's name and address, should be purchased. As a follow-up, it is wise to go online for the purpose of

requesting a *Material Safety Data Sheet* (MSDS), which will include hazardous ingredients, physical data, fire and explosion data, health hazard data, reactivity data, spill or leak procedures, special protection information and special precautions;

4. Store materials safely. Non-breakable, tightly covered containers which are well-labeled should be used; place of storage should be carefully chosen;

5. Have proper ventilation. A window exhaust fan or a local exhaust system that removes contaminated air before it gets into the room are important; simply opening windows and doors to air out the room is not adequate;

6. Wear appropriate personal protective equipment. Since toxic materials can enter the body through the skin and by inhalation as well as by ingestion, gloves and dust masks or air-purifying respirators should be used; safety goggles should also be worn for eye protection; technical data sheets on approved equipment are available;

7. Handle flammable materials safely. Proper containers and disposal procedures should be used; open flames and smoking in the vicinity of flammable materials are extremely hazardous; the proper type of fire extinguisher should be available; electrical equipment should be certified "explosion-proof;"

8. Watch out for physical hazards. Loose clothing, jewelry and hair can get caught in moving machinery; when using sharp tools, cutting should take place away from the body; all electrical equipment should be adequately grounded and the wiring in good repair;

9. Clean up carefully. Vacuuming and wet mopping are preferred to sweeping, which raises dust; chemicals being disposed

of should not be mixed and should be placed in the appropriate disposal container;

10. Use proper personal hygiene precautions. No eating, drinking or smoking should ever take place in the studio; all splashes should be immediately washed off with plenty of water; solvents should never be used for washing up.

If followed, these general rules should be helpful in protecting the artist at work. However, there are particular hazards which are associated with the use of particular art materials and in particular techniques. Table 23.1 summarizes the hazards and precautions that should be taken for the chief areas of artistic endeavor, and the following three sections of this chapter deal with toxic risks involved in using colorants, solvents and other art materials.

23.2. Toxic Colorants

Numerous colorants used in painting, printmaking, ceramics and enameling can cause bodily discomforts with prolonged use, and some can cause long-term effects, damage to vital organs, cancer and even death upon ingestion of small amounts. The chief routes of entry into the body are through the skin, by inhalation and by ingestion. Assuming proper precautions regarding ingestion and skin contact on the part of the artist, the chief route would be through breathing passages to the lungs.

Proper evaluation of the toxicity of colorants would involve measurement of the lethal dose (LD) for ingestion and the time-weighted average Threshold Value (TLV) for inhalation. The LD50 is defined as the dose required to kill 50% of a group of test animals, usually rats; the TLV is defined as the average concentration in air of a substance to which most people can be

TABLE 23.1.

HEALTH HAZARDS IN THE ARTS AND CRAFTS

Technique	Hazard	Precaution
Painting	Acute or chronic heavy metal poisoning on inhalation or ingestion.	Adequate ventilaton; respirator; no eating or smoking in area.
	Irritation, intoxication, damage to organs; possible carcinogenic activity of solvents and paint thinners	Exhaust ventilation and organic vapor respirator for large amounts; dilution ventilation for small amounts
Decorative Arts	Caustic action on skin & respiratory system by dye mordants; bladder cancer from benzidine-type dyes	Dust respirator; gloves; avoid ingestion
	Lead poisoning and corrosive action of hydrofluoric acid; respiratory damage possible from soldeering stained glass	Local exhaust ventilation; dust and fumes respirator; gloves
	Heavy metal dusts and fumes can cause damage to vital organs, central nervous system; acids, silica and asbestos also hazardous in jewelry making	Exhaust ventilation; dust and fumes respirator; protective clothing; damp mopping, not sweeping
Printmaking	Lung and skin irritation and central nervous system damage by solvents	Ventilation; gloves; organic vapor respirator; use of least toxic solvents
	Severe burns and eye damage from acids; chemical pneumonia from acid vapors dissolving in lungs	Acid always added to water when diluting, never the reverse; gloves, goggles, protective clothing, ventilation imperative
	Asbestos-containing talcs can cause cancer	Asbestos-free talcs should be chosen

TABLE 23.1.

HEALTH HAZARDS IN THE ARTS AND CRAFTS (continued)

Technique	Hazard	Precaution
Photo-graphy	Skin irritation, eye damage, lung damage, poisoning from developers; death upon ingestion	Gloves, tongs, goggles, dust respirator when mixing from concentrate; rinse splashes
	Irritation of mucous membranes by vapors of acetic acid from stop bath	Local exhaust ventilation; goggles when mixing from concentrate
	Lung corrosion from sulfur dioxide from fixers	Gloves; local exhaust ventilation
	Skin ulcers and allergies from chromium intensifiers; respiratory problems	Dust mask when mixing; gloves
	Corrosive and highly toxic chemicals damage skin and breathing passages	Gloves; goggles; dust respirator when mixing
Sculpture and Ceramics	Silicosis from silica dust	Dust respirator; damp mopping
	Eye injuries from stone fragments and radiation from welding arcs	Gloves; ultraviolet radiation filter goggles
	Heavy metals from glazes and welding cause skin irritation, allergies, chronic poisoning and possibly cancer	Gloves; welding goggles; dust respirator; protective clothing; wet mopping
	Skin and respiratory irritation; intoxication and allergies from plastic resins and solvents; chronic lung disease from plastics decomposition products and sprayed glazes; chronic respiratory disease from sawdust; poisoning from toxic gases in kiln	Exhaust ventilation; spray booth; dust or organic damp mopping

repeatedly exposed eight hours a day, 40 hours a week without adverse effect. TLV's are important since they are used by OSHA (Occupational Safety and Health Administration) for setting workplace standards, but they are difficult to measure because they require special equipment. Hence, it is helpful to have a quick "rule of thumb" way of determining the relative toxicity of a substance. Michael McCann, in his book, *Artist Beware,* has devised a rating system whereby colorants are categorized as not significantly toxic, slightly toxic, moderately toxic and highly toxic with respect to each of the three entry routes, skin contact, inhalation and ingestion. For space-saving purposes, this text will define three toxicity categories as follows:

I. Not Significantly or Slightly Toxic. A substance is not significantly toxic if it causes toxic effects only by exposure to massive amounts or under very unusual conditions; a substance is slightly toxic if it causes reversible, minor injuries under conditions of normal exposure, or causes allergic effects.

II. Moderately Toxic. A substance is moderately toxic if it can cause minor temporary or permanent damage with either one-time use or repeated usage, if it can result in a major injury or fatality from single or repeated exposure to large amounts, or if allergic reactions are observed in a high percentage of the users exposed.

III. Highly Toxic. A substance is highly toxic if it can cause major permanent damage to vital organs and the nervous system, or fatality, upon a single acute exposure or a long-term exposure to normal amounts, or if a high frequency of severe allergies is observed as a result of exposure to normal amounts.

Two special cases are those substances which may or are known to cause cancer (carcinogens and suspected carcinogens) and those which may cause birth defects by interfering with fetal development (teratogens and suspected teratogens). In addition to the toxicity ratings in the tables that follow, the following key will be used:

* Suspected carcinogen
** Known carcinogen
\+ Suspected teratogen
\+\+ Known teratogen

A fairly significant sampling of artists' colorants is listed in Tables 23.2, 23.3 and 23.4. Table 23.2 contains a list of colorants which fall in category I, that is, not significantly toxic or only slightly toxic. The rating is given on the basis of ingestion and inhalation data for the most part and lacks specificity. For more information about each of the colorants listed, the reader is referred to *Artist Beware*. Information includes the most common name of the colorant, its chemical constitution, its common synonyms, and its *Colour Index* usage number. (The *Colour Index,* first published in 1924 and now in its third edition, is the leading work in the field of colorant classification. It provides a twofold system of numbering colorants based upon usage and chemical constitution, and both are appropriately cross-referenced. C.I. chemical constitution numbers are routinely used by colorists to unambiguously identify a colorant which may be known under a variety of names.)

TABLE 23.2
CATEGORY I: NOT SIGNIFICANTLY TOXIC AND NON-TOXIC COLORANTS

Colorant	Chemical Constitution	Synonyms	C.I. Usage Number
Alizarin Crimson	1,2-Dihydroxy-anthraquinone	Madder Red	Pigment Red 83
Alumina	$Al(OH)_3$; Al_2O_3	Aluminum hydrate; aluminum oxide	Pigment White 24
Barium White	$BaSO_4$	Baryta white Blanc fixe	Pigment White 21 Pigment White 22
Bone Black	Carbon	Frankfurt Black Vine Black Ivory Black	Pigment Black 9
Burnt Sienna	Iron oxides		Pigment Brown 6
Chalk	$CaCO_3$		Pigment White 18
English Red	Iron oxides	Light Red; Indian Red; Mars Red; Mars Violet	Pigment Red 101
Ivory Black	Charred animal bone		Pigment Black 9
Lithopone	$BaSO_4$, ZnO, ZnS		Pigment White 5
Magnesium carbonate	$MgCO_3$	Magnesite	Pigment White 18
Mars Black	Iron oxides		Pigment Black 11
Mars Orange	Iron oxides, Al_2O_3		
Mars Yellow	Iron oxides	Yellow ocher	Pigment Yellow 42
Prussian Blue	Ferric ferrocyanide	Paris Blue; Berlin Blue; Milori Blue; Iron Blue	Pigment Blue 27
Titanium Oxide	TiO_2	Titanium White	Pigment White 6
Ultramarine Blue	Silicate of Na, Al, S		Pigment Blue 29
Zinc White	ZnO	Chinese White	Pigment White 4

Table 23.3 contains a partial list of moderately toxic colorants, and Table 23.4, by far the longest list, is a partial enumeration of highly toxic colorants.

A word of caution for painters should be given here regarding the labeling of tube paints. Presently, many artists' pigments have a variety of common names which are used to identify chemically different pigments. For example, two different manufacturers of artists' pigments use the same name, cerulean blue, for two entirely different colorants, namely barium manganate blue (Pigment Blue 33) and copper phthalocyanine with some added zinc oxide (Pigment Blue 15 + Pigment White 4), respectively. But the common interpretation of "cerulean blue" is cobalt and tin oxide blue (Pigment Blue 35). The latter colorant falls into Category II, whereas the two mentioned previously fall into Category III, and one of them is a suspected carcinogen and a known teratogen. An unsuspecting artist, even after having read the literature on colorant toxicity, could end up choosing a highly toxic colorant without even knowing it. For this reason, a highly desirable practice would be the inclusion of the C.I. usage number on the tubes of the artists' paints or online information regarding the chemical substances present, including Material Safety Data Sheets. At present, the consolidated art supplier, ColArt, is very conscious of safety. The websites of its various subsidiaries such as Winsor and Newton (www.winsornewton.com) and Lefranc et Bourgeois(www.lefranc-bourgeois.com) provide such information, but sometimes one must register an account to access it. In any event, these problems should be part of the practicing, concerned artist's consciousness.

TABLE 23.3
CATEGORY II: MODERATELY TOXIC COLORANTS

Colorant	Chemical Constitution	Synonyms	C.I. Usage Number
Carbon Black**	Carbon	Lamp Black	Pigment Black 6
Cerulean Blue	Cobalt and tin oxides		Pigment Blue 35
Chromium Oxide Green	Chromic oxide	Chrome Green Dingler's Green	Pigment Green 17
Cobalt Blue	Cobalt and aluminum oxides	Thénard's Blue	Pigment Blue 28
Cobalt Green	Cobalt and zinc oxides	Zinc Green Rinmann's Green	Pigment Green 19
Cobalt Violet	Cobalt phosphate		Pigment Violet 14
Cobalt Yellow	Potassium cobaltinitrite	Aureolin	Pigment Yellow 40

TABLE 23.4
CATEGORY III: TOXIC COLORANTS

Colorant	Chemical Constitution	Synonyms	C.I. Usage Number
Antimony White	$BaSO_4$: Sb_2O_3		Pigment White 11
Barium Yellow**	$BaCrO_4$		Pigment Yellow 31
Burnt Umber	Iron oxides; MnO_2		Pigment Brown 7
Cadmium Red*	CdS; CdSe		Pigment Red 108
Cadmium Yellow*	CdS		Pigment Yellow 37
Chrome Green**++	$PbCrO_4$; ferric ferrocyanide	Milori Green	Pigment Green 15
Chrome Yellow**++	$PbCrO_4$		Pigment Yellow 34
Diarylide*++	Coupled derivatives of 3,3' dichlorobenzidine	Benzidine Yellow Benzidine Orange	Pigment Yellow 12
Emerald Green**	Copper acetoarsenite	Paris Green	Pigment Green 21

TABLE 23.4 (continued)
CATEGORY III: TOXIC COLORANTS

Colorant	Chemical Constitution	Synonyms	C.I. Usage Number
Flake White	Basic lead carbonate	Cremnitz White; White Lead	Pigment White 1
Lithol Red*	Derivative of β-Naphthol		Pigment Red 49
Manganese Blue	$BaMnO_4$; $BaSO_4$		Pigment Blue 33
Manganese Violet	Manganese ammonium phosphate		Pigment Violet 16
Molybdate Orange**++	Chromate, sulfate and molybdate of lead		Pigment Red 104
Naples Yellow+	Lead antimonate		Pigment Yellow 41
Phthalocyanine Blue*++	Organic copper compound	Thalo Blue; Monastral Blue	Pigment Blue 15
Raw Umber	Clay with oxides and silicates of iron and manganese	Vandyke Brown	Pigment Brown 7
Red Lake C	Barium Salt of a β-Naphthol Derivative		Pigment Red 53
Scheele's Green**	Copper arsenite		Pigment Green 22
Strontium Yellow**	$SrCrO_4$		Pigment Yellow 22
Talc	Asbestos and free silica		Pigment White 26
Vermilion	HgS	Cinnabar	Pigment Red 106
Zinc Yellow**	$ZnCrO_4$		Pigment Yellow 36

23.3. Toxic Solvents

When toxic solvents are discussed, it should be kept in mind that all solvents must be included within this designation. There is no such thing as a safe solvent, except perhaps, water; some solvents are merely less toxic than others.

A solvent may be defined as any substance which is capable of forming a solution with another substance (see Table 19.1 for some properties of solutions). Most solutions are mixtures of two components called a solute and a solvent; the solvent is usually taken to be the component present in the greater amount. For example, when 5 mL of ethyl alcohol are added to 95 mL of water, the resulting mixture is a solution of ethyl alcohol (solute) in water (solvent). If the ethyl alcohol were present in a greater amount than the water, then the solute-solvent roles would be reversed. Solvents may be solids, liquids or gases. The solvents used by artists are liquid for the most part.

Solvents can damage the body in four ways: (1) They can cause skin disease (dermatitis) by dissolving the natural protective oil barrier on the skin surface. Some solvents are irritating on contact, but others may cause no particular pain even while they are wreaking their damage and perhaps being absorbed through the skin and entering the bloodstream. (2) Solvents can cause irritation of the eyes, nose and throat. All solvents can irritate the sensitive membranes of eyes, nose and throat, even at very low levels. (3) Solvents have a narcotic effect on the nervous system. The narcotic effects of dizziness, sleepiness, nausea, fatigue, drunkenness, staggering gait, headaches, etc. are symptoms of adverse effect on the brain and central nervous system itself. (4) A high degree of exposure to some solvents can cause damage to internal organs. Damage to liver, spleen, kidney, the peripheral nervous system and the heart can be caused by these solvents.

Some solvents have been found to cause cancer; others are suspected of damaging the reproductive systems of both men and women. Some solvents can cause psychological problems such as depression, insomnia, irritability, mental confusion and nervousness.

In addition to their effect on the individual, solvents may also constitute a fire hazard since many of them are highly flammable.

The common solvents fall into nine main groups which we shall discuss in turn.

1.   Alcohols. The alcohols are organic compounds characterized by the presence of the -OH group in the molecule. As a class, they are flammable; they cause irritation, narcosis and dermatitis. The least toxic members of this class are ethyl alcohol (ethanol, grain alcohol, denatured alcohol) and isopropyl alcohol (rubbing alcohol) and should be preferred if possible; one of the most toxic is methyl alcohol (methanol, wood alcohol), which causes severe damage to the central nervous system, blindness, and possible death. Its use should be assiduously avoided.

2.   Aliphatic hydrocarbons. These solvents are compounds which contain the elements of carbon and hydrogen, but do not contain the benzene ring structure. Some members of this class are hexane, gasoline, mineral spirits, naphtha and kerosene. They are all flammable and may cause irritation, dermatitis, narcosis and chemical pneumonia. Hexane, which can also cause permanent central nervous system damage, is the most toxic in this class and should not be used; naphtha is the least toxic in this class.

3.   Aromatic Hydrocarbons. These compounds contain the benzene ring structure. Some examples are benzene (benzol), toluene, the xylenes, coal-tar naphtha, nitrobenzene and styrene. All of them cause central nervous system damage and some of them are carcinogenic; they should be avoided as a class.

4.   Chlorinated Hydrocarbons. These compounds are aliphatic or aromatic hydrocarbons to which one or more chlorine atoms are bonded. Some examples are chloroform, carbon tetrachloride, methyl chloroform (1,1,1-trichloromethane) and perchloroethylene. As a class, they cause central nervous system damage, vital organ damage, and are suspect carcinogens. They should be avoided.

5.   Ketones. The ketones are characterized by the presence of a -C=0 group "sandwiched" between two hydrocarbon groups. The least toxic in this class is acetone, which should be used in preference to the others; methyl butyl ketone (MBK) causes central nervous system damage and should not be used.

6.   Ethers. Ethers are characterized by a -0- atom "sandwiched" between two hydrocarbon groups. Ethers as a class are extremely flammable and many of them form explosive peroxides with the oxygen in the air. They should never be used in theater shops.

7.   Esters. Esters are organic salts, the products of reaction between an alcohol and an acid. Fats and oils are examples of esters, as are oil of wintergreen and aspirin. The least toxic solvent ester is ethyl acetate, the chief ingredient of nail polish remover.

8.   Glycols. Glycols are characterized by the presence of two -OH groups next to one another on the same molecule. Cellosolve is the least toxic in this group.

9.   Miscellaneous Solvents. Turpentine, carbon disulfide, tetra-hydrofuran and dioxane fall into this class. Of these, only turpentine should be used, and even this solvent should be replaced with mineral spirits or odorless paint thinner when possible.

More information about these solvents can be obtained online. An excellent source is the website of the Santa Rita Art League given in the references section.

23.4 Other Toxic Materials

Solvents and colorants are not the only toxic materials used by artists and craftspeople. The acrylics, polyesters, epoxy resins, vinyl polymers, polystyrenes and other plastics all have their toxic effects and appropriate precautions should be taken in their use. Virtually all of the chemicals used in photographic processing have a high degree of toxicity. The acids used in lithography and intaglio printmaking are highly corrosive. The resins used in varnishes can cause allergic effects and dermatitis. No doubt, this list can go on and on and will expand considerably as new techniques and materials are used by artists.

Certain techniques are more hazardous than others also. For example, the Center for Occupational Hazards reports that silk screen printing seems to be one of the most hazardous areas in the arts and crafts due to exposure to massive amounts of solvents containing aromatic hydrocarbons. Therefore, in addition to the normal precautions to be observed in pursuing any art or craft, it is often wise to inquire into safer substitutes; water-based screen printing, therefore, would be preferred to regular silk screen printing if the technique could be perfected to the artist's satisfaction.

In summary, consciousness-raising on the part of educators and responsible, informed practice on the part of artists can go far in reducing the high incidence of occupational diseases and injuries observed among practicing artists and craftspeople.

23.5 Selected Readings

Art Is Creation
http://www.artiscreation.com/index.html
This site contains a comprehensive database on pigments.

410

Association of Occupational and Environmental Clinics
http://www.aoec.org/resources.htm
The next to last entry contains a PowerPoint on the health
hazards of solvents.

McCann, M. *Artist Beware, Updated and Revised: The Hazards
in Working with All Art and Craft Materials and the
Precautions Every Artist and Craftsperson Should Take*;
Lyons Press: Guilford CT, 2005.

McCann, M.; Babin, A. Health Hazards Manual for Artists, 6th
Ed.; Lyons Press: Guilford, CT, 2008.

Santa Rita Art League
http://srart.org/PaintingHazards.html

True Art Information
http://www.trueart.info/health_books.htm
Newsletters and hazard information on many artistic techniques.

U.S. Department of Health and Human Services: Environmental
Health and Toxicology: *Enviro-Health Links* - Keeping the
Artist Safe: Hazards of Arts and Crafts Materials
http://sis.nlm.nih.gov/enviro/arthazards.html

Waller, J.A.; Barazani, G.C. *Safe Practices in the Arts & Crafts:
A Studio Guide*; College Art Association of America: New
York, 1985.

Index